ADVANCES IN NOISE ANALYSIS, MITIGATION AND CONTROL

Edited by **Noor Ahmed**

Advances in Noise Analysis, Mitigation and Control
http://dx.doi.org/10.5772/61648
Edited by Noor Ahmed

Contributors

Thomas Haase, Jiri Pribil, Anna Pribilova, Ivan Frollo, Elisabete Freitas, João Lamas, Carlos Silva, Francisco Soares, Sandra Mouta, Jorge Santos, Min Lei, Yoshiharu Soeta, Antonio Uris, Constanza Rubio, Sergio Castiñeira-Ibáñez, Juan Vicente Sánchez-Pérez, Pilar Candelas, Francisco Belmar, Nikola Stosic, Elvedin Mujic, Ahmed Kovacevic, Ian Smith, Philippe Gaubert, Akinobu Teramoto, Carlos Carvalhais, Joana Santos, Manuela Silva, Ana Xavier, Jean-Francois Brudny

CBS, Edition **2017**

Published by InTech

Janeza Trdine 9, 51000 Rijeka, Croatia

Publishing Process Manager Edi Lipovic

Technical Editor SPi Global

Cover Designer InTech Design Team

Additional hard copies can be obtained from orders@intechopen.com

Advances in Noise Analysis, Mitigation and Control, Edited by Noor Ahmed
p. cm.
Print ISBN 978-953-51-2674-4
Online ISBN 978-953-51-2675-1

INTECH

open science | open minds

Contents

Preface

While rapid industrialisation and urbanisation have underpinned the growth and character of modern day civilisation, they have also given rise to unsustainable energy usage and lifestyles that have the potential to destroy the very environment that supports life (Ahmed & Cameron, Renewable & Sustainable Energy Reviews, vol. 38, pp 439–460, 2014; Ahmed, Procedia Engineering, Elsevier Publications, vol. 49, 2012, pp 338–347). Thus there are many issues that need urgent attention and addressing the significant issue 'noise' from the operation of machines, transport and other man-made systems is one of them.

Today, the term 'noise' has evolved to be more complex than 'unwanted sound', with its definition encompassing random fluctuations of signals originating from a variety of sources, such as human vocal cord, electronic devices, cars, aircrafts, trains, musical instruments and so forth. However, the detrimental effects of some forms of noise and the pace with which they have accelerated in the last few decades are alarming and have the potential to adversely impact our daily activities, health and well-being, prompting intense research to mitigate and control their detrimental effects. This is the background against which work on the present book was undertaken.

This book is a collection of works of eminent researchers from various countries of the world on some selected but important aspects of noise with a heavier emphasis on what can be described as noise pollution. The book is arranged in two sections, each comprising five chapters. For easy identification, the chapters are numbered from one to ten. While there is some overlap of ideas and concepts, the first section, Section 1, broadly comprises chapters that deal with technologies or methodologies to advance the cause of noise analyses, while the second section, Section 2, contains chapters that are more geared to approaches towards noise mitigation and its control, specifically in transport, industry and health sectors.

Section 1: Noise Analysis

It is important to realise that in any research, the capacity to investigate and analyse has been greatly aided by modern computers and their ability to store increasingly greater amount of information, which itself relies heavily on advancements in electronics and their methods of production.

One way to achieve advancement in analysis is by reducing the 'noise' of electronics used in information gathering. Chapter 1 of Section 1, by Gaubert and Teramoto, deals with this aspect. It considers the reduction of noise levels of MOSFET or metal-oxide-semiconductor field-effect transistor, a type of transistor, which is widely used in amplifying or switching electronic signals. The authors show that low resistivity source and drain electrodes can greatly lower the low-frequency noise level of MOSFET by suppressing their contribution to the total noise. They show that new plasma processes having the advantages to work at low electron temperature can also achieve a further reduction through the fabrication of a better gate oxide as well as reduction of damages generally induced by conventional plasma processes. The newly developed MOSFET devices with a lower noise level then conventional structures are, therefore, a serious platform for the next generation of on-board, battery-powered semiconductor chip or 'complementary metal-oxide semiconductor (CMOS)' technology for storing information inside computers.

The following chapter, Chapter 2, considers the critically important aspect of data analysis. Data obtained in experimental situations, especially in real environments, are generally very complicated, chaotic and heavily corrupted by noise. The authors Min Lei and Guang Meng propose a novel approach which they term 'symplectic principal component analysis (SPCA)' to reduce noise of continuous chaotic systems. The authors show that the SPCA method can yield more reliable results for chaotic time series under the different data lengths and sampling times, especially with short data length and under-sampled sampling time, than the classic principal component analysis (PCA).

Chapter 3 by Thomas Hasse, entitled 'Optimal placement of sensors and actuators for feedforward noise and vibration control', contains a detailed analyses of parameters such as the causality, filter weights and the damping properties that have significant influence on the performance of a feedforward control system. Studies of the past three decades suggest that to devise any successful practical system to reduce noise and vibration, it is imperative to look into the locations of detectors and actuators and how they are arranged. The author in this chapter demonstrates that a prior knowledge of the influences of these parameters should be established that will result in the optimisation of the location of detectors and actuators for better performance.

Chapter 4 discusses variable drive of AC systems which are used in electromechanical systems that can range from small appliances to large mine mill drives and compressors and are integral parts of modern day life. The authors of this chapter, Brudny, Morganti and Lecointe, describe control strategies to reduce the noise of magnetic origin through the rejection at high frequencies of the voltage switching 'three-phase harmonic systems' (STPHSs) using a strategy called 'carrier-phase jump'. Their proposed calculation method is simple to implement and characterises STPHS phase sequences that have a direct impact on the definition of the force components at the origin of the magnetic noise generated by AC machines and subsequent control strategies for reducing this noise.

In Chapter 5, the final chapter of Section 1, Freitas, Lamas, Silva, Soares, Mouta and Santos detail a new methodology to assess the 'annoyance' aspect of road-traffic noise that predominantly occurs at speeds above 40 km/h. The negative impact of 'annoyance' on health is generally mitigated by optimising road pavement surface characteristics. However, experimental data linking surface characteristics and annoyance rate remains scarce. Moreover, because of the complexities of assessing annoyance by experimental means using real sounds, the authors have resorted to developing virtual sounds that have the real possibility to accurately establish requirements to control traffic noise based on indicators that better describe the annoyance.

Section 2: Noise Mitigation and Control

Section 2 contains advances in methods to control and mitigate the effects of noise. To maintain continuity with Section 1, the five chapters of Section 2 are numbered from Chapters 6 to 10. The first chapter of this section—Chapter 6—describes an approach that is based on the postulation that if noise cannot be reduced at source, then attempts to mitigate it during its transmission phase become appropriate. Consequently, for the transmission phase, the authors, Rubio, Ibariez, Perez, Candelas, Belmar and Uris, develop the concept of 'open acoustic screens'. These screens, or the 'subwavelength slit acoustic screens' as the authors call them, employ a new type of attenuation mechanism that are not based on crystalline geometry but consist of arrangements of isolated scatterers. Such acoustic screens have interesting properties and can be considered as real alternatives to the classical screens formed by continuum rigid materials.

Noise reduction in underground stations using platform screen doors (PCD) forms the basis of Chapter 7. Underground stations are now an integral part of urban city transport systems of most parts of the world. The authors, Soeta and Kim, use both computer simulation and scale model testing of several PCDs of different configurations and sizes. They found that speech intelligibility and the sound pressure level were increased by most types of PSDs. These findings are significant and pave the way for the development of better public announcement systems in trains which will give safety and comfort in hearing of passengers.

Chapter 8, by Mujic, Kovacevic, Stosic and Smith, identifies gas pulsations as the most important source of noise in screw compressors that can use a variety of fluids and are widely used in a number of industrial and domestic applications. Using both numerical modelling and physical tests on industrial screw compressors, the authors demonstrate that the gas pulsations in the compressors and the consequent noise that they generate can be substantially reduced through appropriate changes to the shape of the compressor discharge port.

The last two chapters of section 2 tackle problems of noise that bear direct relations to the medical world. The authors, Pribil, Pribilova and Frollo, of Chapter 9 focus their attention on improving the performance and application of magnetic resonance imaging (MRI). The discovery and development of magnetic resonance imaging (MRI) scanners have revolutionised medical procedures and research around the world. The authors describe methods to reduce noise of the speech recorded in an open air MRI working in a weak magnetic field during human phonation for the vocal tract modelling. The achieved results will prove useful as a database in the design of filters for noise suppression in the speech signal recorded simultaneously with three-dimensional human vocal tract scanning.

The final chapter, or Chapter 10, is a 'neo-project' on noise reduction and control in a hospital environment. The authors, Carvalhais, Silva, Xavier and Santos, provide the groundwork for an integrated approach to minimise sound pressure levels in neonatal intensive care units where exposure to sound pressure levels can significantly influence the quality and well-being of the workers and others, as well as the recovery of premature infants who are hospitalised. The outcomes of this project will contribute towards educating health-care staff and providing good practice guide in a daily basis and reduce and control noise production in hospital environments.

This book is intended to be a valuable addition to students, engineers, scientists, industrialists, consultants and others by providing greater insights into noise-related problems, particularly in noise pollution, and methodologies to study, mitigate or control them.

All chapters have been prepared by professionals who are experts in their respective research fields and the contents reflect their views. All chapters included in this book are a culmination of the interactions of the editor, publisher and authors.

The editor takes the opportunity to thank all the authors for their expert contributions and the publisher for their patience and the hard work that has gone in producing this book of high practical significance.

Noor Ahmed, BSc (Hons), PhD, CEng, Fellow IEF, Fellow IMechE
Department of Mechanical Engineering Technology within the Faculty of Engineering and the Built Environment
Auckland Park Kingsway Campus
University of Johannesburg, South Africa

Section 1

Noise Analysis

Chapter 1

New Processes and Technologies to Reduce the Low-Frequency Noise of Digital and Analog Circuits

Philippe Gaubert and Akinobu Teramoto

Additional information is available at the end of the chapter

http://dx.doi.org/10.5772/64630

Abstract

The chapter is intended to provide the reader with means to reduce low-frequency noise in Metal-Oxide-Semiconductor Field-Effect-Transistor (MOSFET). It is demonstrated that low-resistivity source and drain electrodes can greatly lower the low-frequency noise level by suppressing their contribution to the total noise. Furthermore, new plasma processes having the advantages to work at low electron temperature can achieve a further reduction, thanks to the fabrication of a better gate oxide and to a reduction of damages generally induced by conventional plasma processes. Reducing the impact of the traps on the carrier flowing inside the channel by burying the channel can also achieve a reduction of the noise level, but unfortunately at the cost of a degradation of the electrical performances. Finally, the noise analysis of the low-frequency noise in accumulation-mode MOSFETs showed that these newly developed devices have a lower noise level than conventional structures, which, in addition to their superiority in term of electrical performances, establishes them as a serious platform for the next Complementary Metal-Oxide-Semiconductor Field-Effect-Transistor (CMOS) technology.

Keywords: low-frequency noise, radical oxidation, silicide, series resistances, accumulation-mode, MOSFET, buried-channel, fabrication process

1. Introduction

Since the dawn of electronics more than 50 years ago, manufacturers have been providing customers with faster and smaller chips by fabricating increasingly better devices and improving processes. The main strategy adopted has been to shrink the gate size of the MOSFETs to improve chip performances, especially speed. Since the signal-to-noise ratio was large enough, the noise was not an issue and its reduction dragged very little effort among the scientific community. After a steady working frequency doubling every year, the recent downscaling of the dimension has led to high stress and increased variability and, in turn, stagnation of performances of chips. There is no doubt that the increased noise is to blame for that standstill, even if other problems such as the doping concentration could be implicated as well. Nevertheless, the distinction between noise and signal has become critical and the noise issue must be tackled. A suppressed noise level should pave the way to once again lower biases leading to less heat, better reliability, and better performances.

Noise is a fluctuation of a quantity that shifts back and forth with uncertainty. In electronics, it is generally noted as a fluctuation of the voltage or current around its mean value and is ascribed to stochastic events which find their origin at a microscopic level trough the discrete nature of the transport or the Brownian nature of the carrier. There are several types of noises such as thermal noise, shot noise, generation-recombination noise, inter-band noise, and low-frequency noise. They are generally classified upon their origin. Among these, the thermal noise and low-frequency noise are of paramount importance in MOSFETs, with the latter one being of most concern since its origin is still in debate and that its evaluation for a given technology is made extremely difficult. Even if the low-frequency noise has been a limiting factor of performances for analog circuits for several years, it has recently become as well an issue for digital ones. Indeed, even though its limitation applies in the low range of frequency, it is up-converted into phase noise leading to time domain instabilities and therefore problems in the high-frequency range. Its reduction is therefore mandatory not only for analog but also for digital circuits.

In Section 1, the theory of the low-frequency noise in MOSFETs is briefly reviewed. While Sections 2 and 3 present new technologies to suppress it by the means of, respectively, silicide and damage free processes, Sections 4 and 5 introduce improved MOSFETs. Thus, the results regarding buried-channel and accumulation-mode MOSFETs are reported, respectively, in Sections 4 and 5.

2. Low-frequency noise in MOSFETs

The MOSFET is a complex device composed of purely resistive parts surrounding the channel whose resistance is controlled by the bias applied at the gate electrode. It is therefore natural that the noise generated inside each region is propagating up to the source and the drain electrodes. However, the noise stemming from the channel is generally the most dominant one, even though the one coming from the surrounding areas can play an important role and can even take over as the main noise source [1]. **Figure 1(a)** shows a schematic of a MOSFET structure while **Figure 1(b)** represents its equivalent noise circuit. The source access resistance, the drain access resistance, and the channel are the three main regions the noise is coming from and the total measured noise S_{Id} at a given frequency can be written [2].

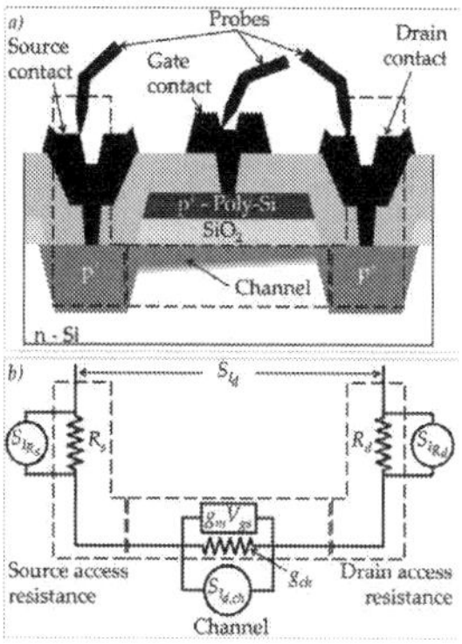

Figure 1. Representation of (a) MOSFET and (b) its equivalent circuit noise.

$$S_{I_d}(f) = \frac{S_{I_{d,ch}}(f)}{\left[1 + g_m R_s + g_{ch}(R_s + R_d)\right]^2} + \frac{g_m^2 R_s^2 S_{I_{R_s}}(f) + g_{ch}^2 \left[R_s^2 S_{I_{R_s}}(f) + R_d^2 S_{I_{R_d}}(f)\right]}{\left[1 + g_m R_s + g_{ch}(R_s + R_d)\right]^2} \tag{1}$$

where $S_{Id,ch}$, $S_{IRs,}$ and S_{IRd} are the noises generated, respectively, inside the channel, the source R_s and drain R_d access resistances. G_m and g_{ch} are, respectively, the transconductance and conductance of the channel. The left-hand side term is the contribution coming from the channel while the right-hand side one is the contribution coming from the access resistances.

It is worth noticing that S_{Id} can become equal to $S_{Id,ch}$ meaning that the measured noise will be the noise of the channel, whereas even if the series resistances are the main noise source, it is their contribution that will be measured and not their pure noise. The noise in a resistive material such as the access resistances of a MOSFET is either thermal noise or $1/f$ noise at low frequencies [3]. This $1/f$ noise is known to originate from the fluctuations of the fundamental mobility and it follows the Hooge's empirical formulae [3]:

$$\frac{S_I(f)}{I^2} = \frac{\alpha_H}{fN}, \tag{2}$$

which represents the normalized noise of the current I while N is the number of carriers. α_H is the Hooge parameter suggested to be constant and to reflect the quality of the crystal. After several decades of controversies between the Hooge theory and the Mc. Whorter one [4], the latter explaining the noise in terms of fluctuation of the carrier number, Mikoshiba [5] developed a new theory able to gather all data into a single model to explain the noise stemming from the channel.

His theory has been confirmed afterward by several researchers and is now well accepted among the scientific community. Within this theory, the noise is given by [6]

$$\frac{S_{I_d}(f)}{I_d^2} = \frac{g_m^2}{I_d^2}\left(1 \pm \alpha\mu_{\mathrm{eff}}C_{\mathrm{ox}}\frac{I_d}{g_m}\right)^2 S_{V_{fb}}(f), \tag{3}$$

where C_{ox} is the gate oxide capacitance, g_m is the transconductance, and μ_{eff} is the effective mobility of the carriers flowing inside the channel. This theory is known as the insulator and induced mobility fluctuations, and is ascribing the origin of the noise to the traps located inside the gate insulator near the interface. The constant dynamic capture and release of carriers from and to traps generate interfacial insulator charge variation and, in turn, fluctuation of the insulator charge. These fluctuations are equivalently generating flat band variations. These fluctuations are summarized on the left-hand side term in Eq. (3) and are proportional to the flat band voltage fluctuations S_{Vfb}, expressed as [6]

$$S_{V_{fb}}(f) = \frac{\lambda k_B T q^2 N_t}{WLC_{\mathrm{ox}}^2 f}, \tag{4}$$

with k_B being the Boltzmann constant, q the electron charge, T the temperature, and λ the tunnel attenuation length of the traps in the insulator equal to 0.1 nm for SiO_2. N_t is the interface trap density. In addition to the variation of the insulator charge, the capture and release mechanism is locally affecting the surface potential at the interface, resulting in a Coulomb interaction between the locally deformed potential surface and the carriers flowing inside the channel. The localized scattering rate will vary and will induce fluctuations of the mobility. These fluctuations are ascribed to the right-hand side term in Eq. (3) and are related to the fluctuations of the insulator charge through the Coulomb parameter α, which measures the strength between both quantities. With nowadays miniaturization of the gate of the MOSFETs, two kinds of noises are at stake, although they are both explained in the frame of the previous theory. Indeed, it is obvious that in the case of very small gate size involving a very limited number of carriers, the removal or introduction of a single free carrier scared within the total number of free carriers involved in the conduction will have a tremendous impact on the current, making it jump between two or several levels like randomly disposed crenels. This type of noise is called random telegraph noise and is commonly an issue for sensors, especially optical ones [7]. However, when the number of traps is more significant and for a specific distribution of their energy within the bandgap, the resulting noise is commonly called $1/f$ or even Flicker noise and it follows a distribution, proportional to the inverse of the frequency when plotted as a function of the frequency [1]. This noise is an issue for analog circuits, and even for some digital ones, and will even impact at high frequency due to its conversion into phase noise. Finally, the noise in MOSFETs can be summarized as depicted in **Figure 2**. The noise measured at the electrode of a MOSFET is the sum of three terms: the fluctuation of the insulator charge, induced fluctuation of the mobility, and the contribution coming from the access resistances. It is worth mentioning that there is a fourth term, the cross-correlation term between the fluctuation of the insulator charge and the induced mobility one even though it does not have a physical origin.

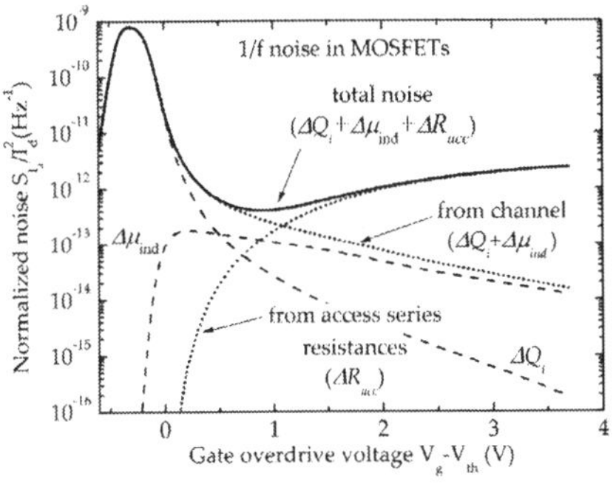

Figure 2. Normalized noise in a MOSFET. The several contributions of each noise sources have been reported with the non-full lines.

3. Source and drain contacts

When it comes to low-frequency noise, the contribution stemming from the source and drain access series resistances is generally overlooked. This negligence can have tremendous impact on the noise analysis especially at high gate voltage where their contribution will mostly take over as the dominant noise.

As a matter of fact, this is not the noise generated inside the source and drain access series resistances, which is at stake but their contribution. Rather than reducing the noise sources, reducing their contribution is more efficient and easier. Indeed, the reduction of the resistance of the source and drain access contacts does not only mean a better drivability and a better transconductance but can guarantee a reduction of the propagation of the noise generated by the noise sources inside the contacts and, in turn, a reduction of the contribution of the source and drain contact noise to the total measured noise [8]. The performance improvement of CMOS has become of paramount importance with scaled dimension. Much effort is being made to increase the carrier mobility by several means such as strained technology [9], different silicon orientation [10], or even different semiconductor [11]. The reduction of the source and drain electrode series resistances is another means to improve the drivability and silicide has already been used for such purposes. Nitride alloy silicide is widely used to lower the Schottky barrier height to either n^+ or p^+ silicon down with contact resistance as low as $2 \times 10^{-9}\ \Omega\ cm^2$ in the best case [12]. A new structure [13] and new processes [14] have been developed in order to further lower down the series resistances. Instead of using the same silicide for both p^- and n-MOSFETs, erbium has been selected to perfectly fit the requirement of n-MOSFETs and palladium for the p-MOSFETs. Additionally, tungsten metal stack above the thin silicide layer was introduced to reduce the sheet resistance and protect erbium from being oxidized. In order to confirm the above statement, two kinds of MOSFETs have been fabricated following the very same process flow, except during the source and drain contact fabrication stage. The source and drain contacts of the reference transistors have been fabricated with aluminum (Al), while erbium silicide associated with tungsten ($ErSi_2$/W) has been used for the second set of transistors. The respective structures have been represented in **Figure 3(a)** and **(b)**. While both wafers followed the same process flow until the contact lithography step, the $ErSi_x$/W wafer followed an advanced process entirely developed at New Industry Creation Hatchery Center (NICHe). In this advanced process, the wafer has been loaded in an N_2 sealed cleaning chamber after a total room temperature five-step cleaning, which has been followed by the dipping of the cleaned wafer in O_3 dissolved ultra pure water in order to form a chemical oxide at the silicon surface. The removal of the chemical oxide has been carried out by diluted HF solution and the wafer has been then transferred in clustered sputter equipment, still in N_2 ambient, where the formation of a thin film of erbium followed by the deposition of a tungsten capping layer has been done by Radio Frequency (RF) sputtering.

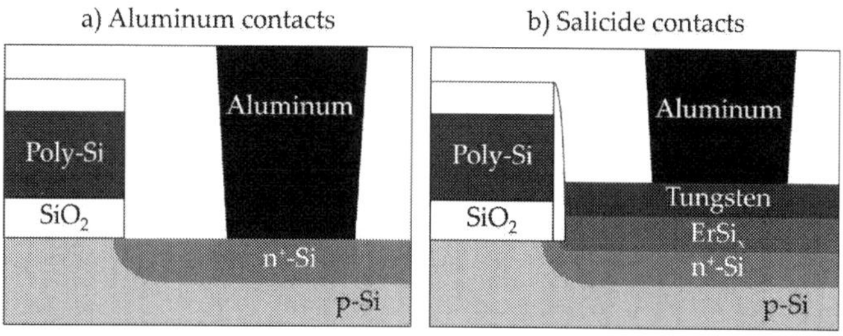

Figure 3. Schematic of the structure of (a) reference contacts fabricated with aluminum (Al) and (b) salicide contacts fabricated with erbium silicide ($ErSi_x$) and tungsten (W) layer. Copyright 2011 The Japan Society of Applied Physics [8].

The wafer has been then loaded in lamp annealing chamber to finally form $ErSi_x$. The wafer has been then brought back to the conventional process flow to finally form aluminum contacts. Electric characterization has been carried out and the main results are summarized in **Figure 4(a)** and **(b)**. As expected, the drivability has been improved by a factor of 2 on account of silicide contacts [8]. Furthermore, the maximum of the transconductance also increased and confirms the interest of low-resistivity source and drain contacts to enhance performances of electronic circuits. Noise measurements have been performed in the linear regime and for f = 10 Hz. The result is presented in **Figure 5**. When compared with the noise level of the reference transistor, the noise level of the transistor featuring $ErSi_x$/W contacts is greatly reduced for positive gate overdrive voltages while it remains equivalent for lower values. The same noise level below 0 V is explained by the fact that within this range of measurement the channel is exclusively contributing to the total noise and by the fact that both devices have indeed almost the same channel, since they followed the same process flow with regard to the fabrication of the gate stack. The noise modeling has been carried out and is reported with the lines in **Figure 5**. The modeling reveals that from 0 V the noise from the reference device moves away from the noise stemming from the channel. The contribution of the series resistances to the total noise is increasing and is ultimately taking over as the main contribution. However, the noise of the device featuring $ErSi_2$/W contacts is following the curve depicting the noise stemming from the channel and starts to slightly move away at high voltages. The impact of the series resistances on the noise is barely visible. The use of low-resistivity contacts allows a drastic reduction in the contribution of the series resistances to the total noise and let the channel be the sole source of noise over the entire measurement range [8]. About 10 times reduction of the access resistance has led to 100 times reduction of the noise level.

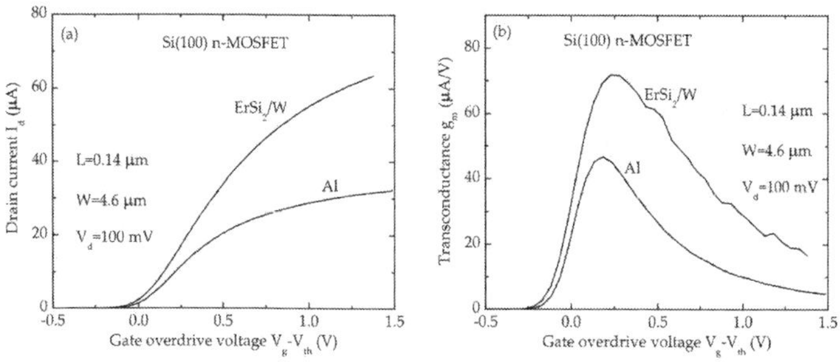

Figure 4. (a) Drain current and (b) transconductance versus gate overdrive voltage for n-MOSFETs featuring Al and $ErSi_2$/W contacts. Copyright 2011 The Japan Society of Applied Physics [8].

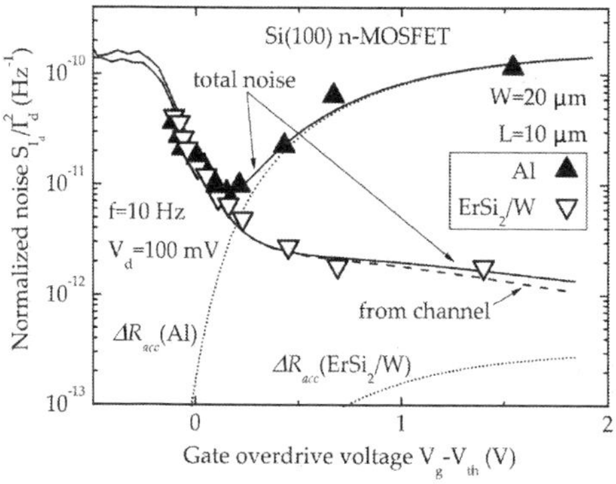

Figure 5. Normalized noise of n-MOSFETs with contacts fabricated with either aluminum(Al) or erbium silicide/tungsten($ErSi_2$/W). (W) layer. Copyright 2011 The Japan Society of Applied Physics [8].

4. Radical oxidation

The gate stack, especially, the gate insulator, is the most critical part of the MOSFET, mainly because of the defects that can appear during the fabrication and its tremendous impact on the device performance [15]. It is absolutely true these days that the need of always-faster devices and smaller chips also promote the appearance of undesirable effects such as increase of variability and random telegraph noise. Thermal oxidation has been the way, since the establishment of the MOSFET, to fabricate the gate insulator and while the generated SiO_2 was at the beginning of poor quality, leading to high *S* parameters, the process has greatly evolved since and the growth of quality oxide can be achieved now [15]. Unfortunately, the dimension of nowadays MOSFETs has reached a threshold where *S* parameter and V_{th} variability are a major issue, as well as, noise level prevision [16].

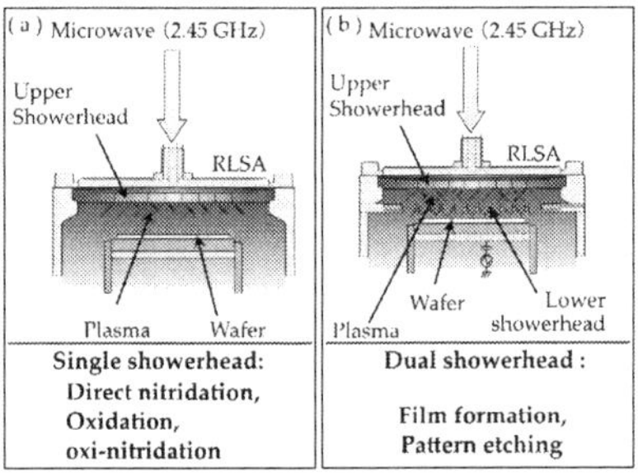

Figure 6. Diagrams of (a) single shower plate and (b) double shower plate equipment based radicals formed by the low electron temperature microwave high-density plasma.

Thermal oxidation, from its intrinsic chemical reaction, cannot be optimized anymore and will always promote the formation of damage (either inside the gate stack or cap layers) and will partly invalidate the flattening process and, in turn, deteriorate the surface roughness of the wafer. Thus, new oxidation processes have been developed to avoid these issues. They are all based on radical oxidation rather than chemical reaction to form SiO_2 [17]. The specificity of the damage-free very low electron temperature microwave exited high-density plasma is, as represented in **Figure 6**, that it can be employed for oxidation at low temperature, chemical vapor deposition, or even reactive ion etching. Very high quality gate insulator and reduced damages generally occurring during the etching and the fabrication of interconnect can be achieved thanks to this advanced process as shown in **Figure 7** [15, 18, 19]. Contrary to the thermal oxidation, the radical oxidation has an oxidation rate that is almost regardless of the orientation of the silicon crystal on which the oxide is grown [20]. Additionally, the radical oxidation does not only help reduce the interface trap density but also help preserve and even improve the flatness of the Si/SiO_2 interface. Two sets of p-MOSFETs have been fabricated in our clean room. They followed almost the same process flow; however, they differed in such a way that the first set featuring a radically grown oxide has been processed exclusively with

http://dx.doi.org/10.5772/64630

advanced plasma equipment, while the second set has been processed using conventional processes, among which the thermal oxidation process.

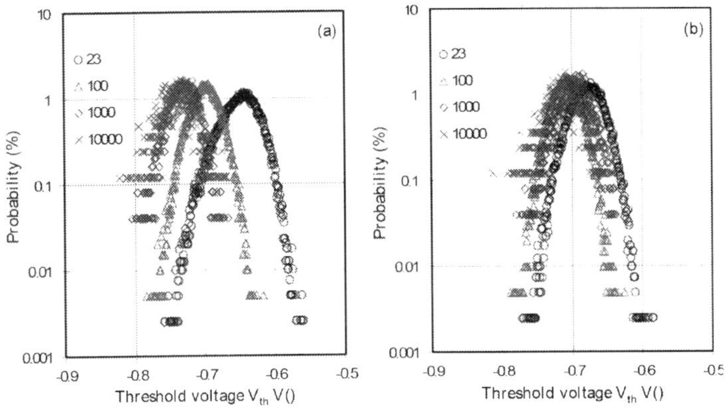

Figure 7. V_{th} distributions of p-MOSFETs fabricated by applying (a) conventional plasma processes and (b) radial line slot antenna plasma processes plotted for antenna ratio of 23, 100, 1000, and 10,000.

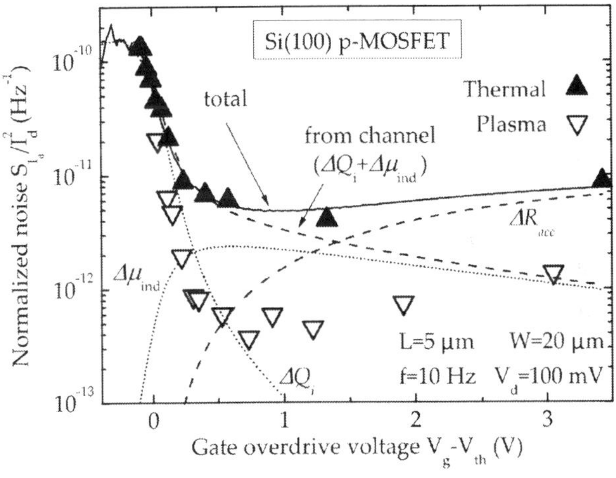

Figure 8. Normalized noise of p-MOSFETs with a gate oxide fabricated by either thermal or plasma oxidation as a function of the gate overdrive voltage. The modeling, reported with the lines, refers exclusively to the transistor with a gate oxide thermally grown. Copyright 2011 The Japan Society of Applied Physics [8].

Noise measurements have been performed in the linear and saturated region and for different gate sizes. The noise analysis has been carried out at 10 Hz. Results are presented in **Figure 8**, and they clearly indicate that the p-MOSFET with a gate oxide fabricated by radical oxidation has a lower noise level than when the thermal oxidation process is used used, with a maximal reduction of over a decade. As expected, the noise stemming from series resistances and the noise from the channel are both contributing to the total noise, with the latter one being ascribed to the insulator charge and induced mobility fluctuations. In order to understand the origin of the noise reduction, the modeling of the p-MOSFETs, featuring a gate oxide, fabricated by radical oxidation has been carried out. The result is reported in **Figure 9** and it revealed an unexpected behavior, i.e., no induced mobility fluctuations. The contribution of the series resistances added to the sole insulator charge fluctuations has been enough to model the total noise. Even though the trapping/release mechanism at the origin of the 1/*f* noise induces fluctuation of the mobility, these fluctuations are, in the present case, too small to be visible when compared to the other fluctuations. The interface trap density has been extracted for both sets of devices and revealed a three-time reduction in favor of the transistors fabricated using plasma processes with $N_d = 2 \times 10^{16}$ cm^{-3} eV^{-1}, testifying the high integrity of the oxide when fabricated by radical oxidation.

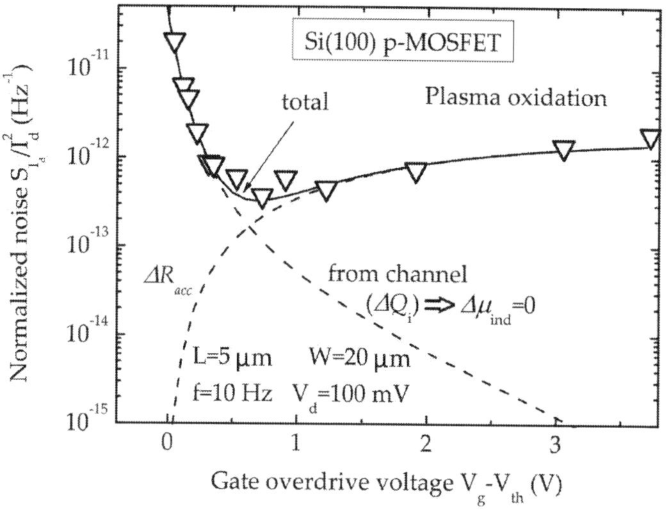

Figure 9. Normalized noise of p-MOSFETs with a gate oxide fabricated by plasma oxidation as a function of the gate overdrive voltage. Copyright 2011 The Japan Society of Applied Physics [8].

5. Buried-channel MOSFETs

5.1. Structure of buried-channel MOSFET

The low-frequency noise, such as 1/*f* noise and random telegraph noise, is basically caused by the defects at the gate insulator and silicon interface in MOSFETs. To reduce the low-frequency

http://dx.doi.org/10.5772/64630

noise, the number of defects or the influence of the defects has to be reduced. Here, the channel of the buried-channel MOSFETs is separated from the gate insulator/Si interface, and then the carriers in the channel are hard to be influenced by the defects at the interface. Usually, the buried-channel MOSFETs are characterized as high mobility but weak short channel Field-Effect-Transistor (FET) [21–23] because the carriers flow through the bulk. In this case, the separated channel location of buried-channel MOSFETs is very important for reducing the low-frequency noise. **Figure 10** shows the schematic illustration of the buried-channel n-MOSFET. The buried layer was formed by the ion implantation at the V_{th} adjustment process [24, 25]. **Figure 11** shows the band diagram of the conventional surface channel (a) and the buried-channel MOSFETs (b), respectively [25]. The buried layer depth was set as 170 nm from the SiO_2/Si interface, and the channel was appeared at around 30 nm from the SiO_2/Si interface at the bias conditions of back bias (V_{BS}) =1.5 V and drain current (I_{DS}) =100 nA for a gate length (L) of 0.22 μm and a gate width (W) of 0.28 μm MOSFETs.

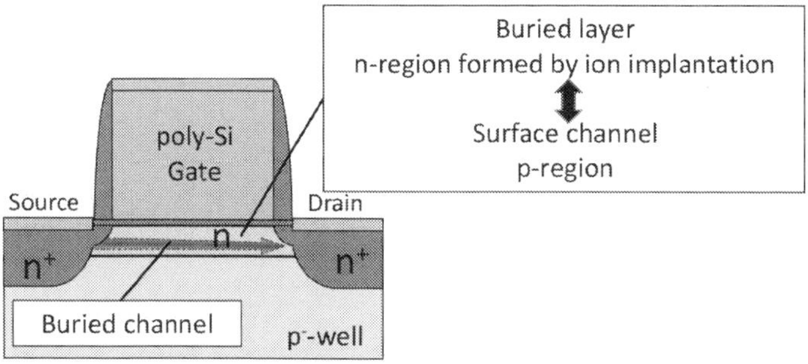

Figure 10. Schematic illustration of the buried-channel MOSFET.

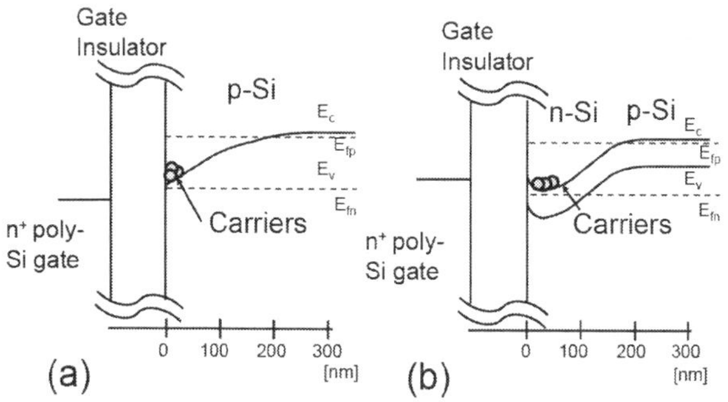

Figure 11. Diagrams of surface channel (a) and buried-channel MOSFETs. The bias condition was set at V_{BS} = 1.5 V and I_{DS} = 100 nA.

5.2. Low-frequency noise characteristics

Figure 12 shows the 1/*f* noise characteristics for the surface channel and buried-channel MOSFETs. Five samples were measured for each MOSFET. The size (*W*/*L*) of MOSFETs was 10 μm/5 μm and bias conditions were set as V_{DS} = 1.5 V, V_{BS} = 0 V, I_{DS} = 1, 10, and 100 μA. S_{Id} increases as the drain current increasing for both MOSFETs. The noise power S_{Id} of the surface channel MOSFETs is proportional to 1/*f*. In contrast, S_{Id} of the buried-channel MOSFETs for the low I_d cases is not proportional to 1/*f*, because the noise level was smaller than the floor noise of the measurement system. For the same drain current I_d, the noise power of the buried-channel MOSFETs are less than that of the surface channel MOSFETs, especially their differences are observed for the low drain current cases. In this experiment, the gate voltage controlled the drain current. When the gate voltage increases, the distance between the channel and the interface decreases, and then it influences the defects. This indicates that the noise reduction of buried channel is very effective for the low gate voltage conditions.

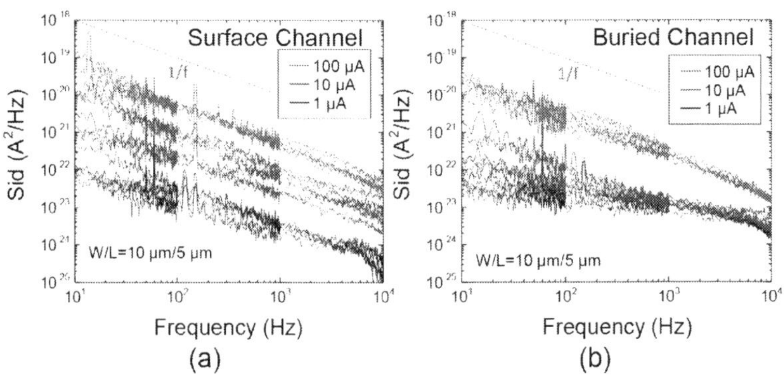

Figure 12. 1/*f* noise characteristics of (a) surface channel and (b) buried-channel MOSFETs.

5.3. V_{th} variability

In the previous section, it is described that the noise can be reduced by introducing the buried-channel MOSFETs because the channel is separated from the SiO_2/Si interface. However, this means the gate capacitance becomes lower compared to the surface channel MOSFETs. Then, the V_{th} control becomes difficult compared with the surface channel. **Figure 13** shows distributions of (a) V_{th} and (b) the channel charge (Q_{ch}) with 65536 MOSFETs for the buried- and surface channel MOSFETs [24]. These distributions are measured by the array test circuit, which can measure the V_{th} variation and the random telegraph signal for many MOSFETs (>1 million MOSFETs) during very short time (<1 s) [26–29]. The horizontal axis of both the graphs shows the difference values between the average values of all MOSFETs (65536 MOSFETs). The V_{th} variability of the buried-channel MOSFET is larger than that the surface channel MOSFETs as shown in **Figure 13(a)**. It is considered that the gate capacitance of buried channel is smaller than that of the surface channel. Then, the horizontal axis is converted from V_{th} to

Q_{ch} by using the gate-channel capacitance. Almost the same distributions are observed, and then the variability is caused by the small capacitance between the gate and the channel. Moreover, it is noticed that the noise increases by the excess capacitance decrease with the other transistor characteristics degradation, such as short channel effect and subthreshold swing degradation [22, 30].

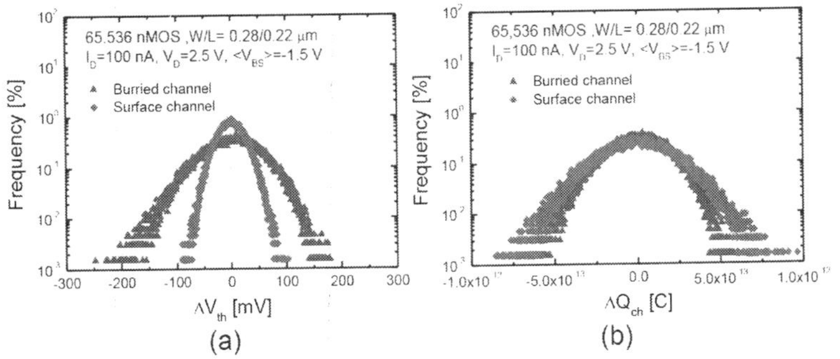

Figure 13. Distributions of (a) V_{th} and (b) Q_{ch} with 65536 MOSFETs for the buried- and surface channel MOSFETs.

6. Accumulation-mode MOSFETs

The separation between the interface and the channel is effective in reducing the noise in buried-channel MOSFETs; however, the controllability of V_{th} is worse than conventional MOSFETs because the gate capacitance is reduced. Then the introduction of Silicon-on-Insulator (SOI) wafer and a new structure can solve this issue [31, 32]. The so-called accumulation-mode MOSFET has been developed keeping this in mind. As depicted in **Figure 14**, the accumulation-mode MOSFETs differ from MOSFETs in such a way that the type of the SOI layer is the same as that of the contact. Additionally, the type of polysilicon must also be adjusted as required. Although the working of the conventional inversion-mode MOSFETs is based on the generation of an inversion layer made of the minority carrier, the accumulation-mode one is making use of an accumulation layer composed of majority carrier. Actually, the accumulation-mode MOSFET without bias is at first on the off-state since the SOI layer is completely depleted. When a bias is applied at the gate, a thin conductive layer of majority carrier is first generated at the back interface between the Buried Oxide (BOX) and the SOI. Then accumulation-mode MOSFETs become on the on-state. A current at the back interface flows from the source to the drain. A further increase in the bias makes this layer disappear and makes a short portion of the originally depleted SOI layer become neutral. A bulk current is generated on the BOX side. This current continues to increase with the expansion of the neutral region (due to the shrinking of the depleted region) inside the SOI until the bias applied at the gate reaches the flat-band voltage.

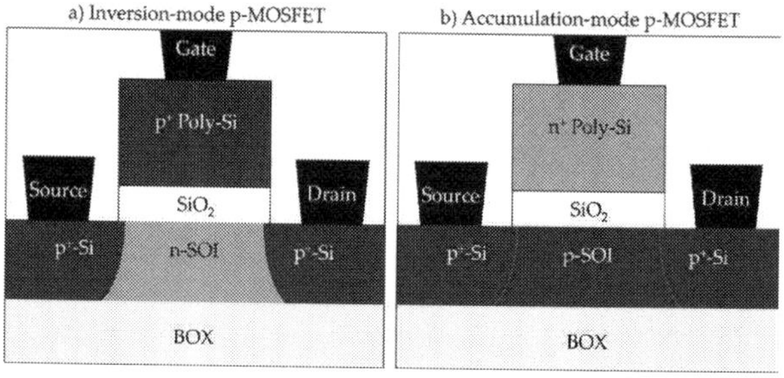

Figure 14. Schematic representation of (a) inversion-mode and (b) accumulation-mode p-MOSFETs.

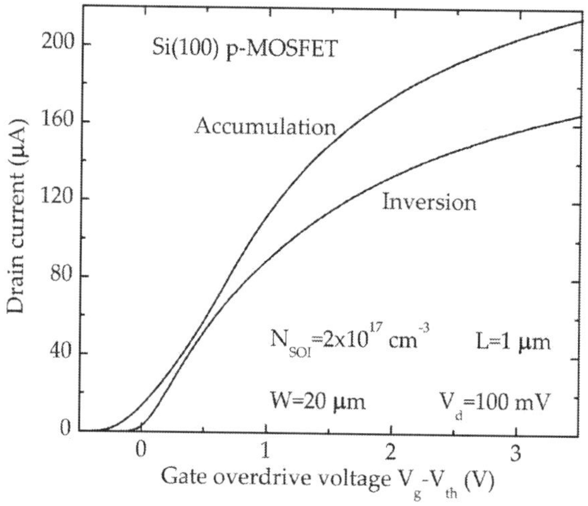

Figure 15. Drivability of accumulation- and inversion-mode p-MOSFETs.

The current generated inside the accumulation layer then adds to the bulk current. Therefore, in addition to, among other things, having an improved reliability [33] and being immune to radiation effects [34], they also have a better drivability than inversion-mode MOSFETs since the total current is the sum of those generated inside the SOI and the accumulation layer [35]. When the bulk current has reached its maximum, corresponding to the SOI completely neutral, the majority carrier accumulates at the front interface between the gate insulator and the SOI. Accumulation-mode fully depleted SOI MOSFETs have been fabricated on Si(100) surface to investigate the noise characteristics. The SOI layer impurity has been adjusted by ion implantation to 2×10^{17} cm^{-3}. The thickness of the SOI layer has been reduced down to 50 nm. In order to avoid the increase of noise due to the defects at the front interface and the impact of the surface roughness of the interface, the radical oxidation [17] added to the five-step-cleaning process [20] has been repeated four times until reaching a flattened interface with a roughness R_a of 0.08 nm. A 7-nm gate oxide has been formed by radical oxidation using a low electron temperature microwave high-density plasma process at 400°C. As expected and as shown in **Figure 15**, the accumulation-mode MOSFETs feature a better drivability than those of the inversion-mode MOSFETs. The better drivability does not only owe to two distinctive currents but also from the higher carrier mobility thanks to a lower transversal electric field at the front interface [35].

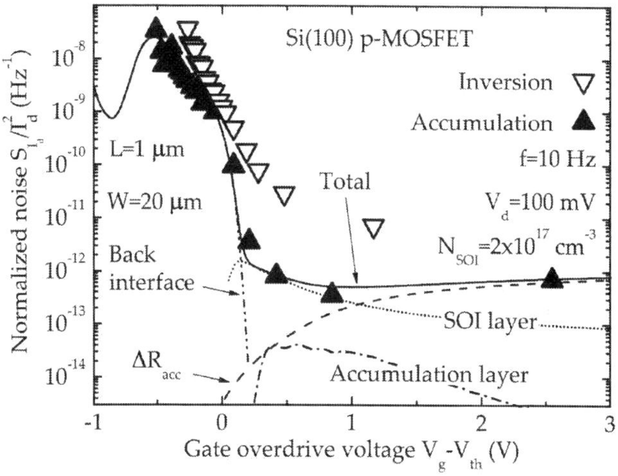

Figure 16. Normalized noise of accumulation- and inversion-mode p-MOSFETs. Lines refer to the modeling of the noise of the sole accumulation-mode device and represent the noise generated by each region.

The noise of the inversion- and accumulation-mode MOSFETs has been reported in **Figure 16**. Even though their noise level is similar at low gate overdrive voltage, the superiority of the accumulation-mode MOSFET over the inversion mode one becomes clear for voltage over 0 V. A gap of more than 1 decade is achieved at the best. While the noise of the inversion-mode MOSFET can be ascribed to the interface traps [6] and series resistances at high bias [3], the origin of the noise for the accumulation-mode one must be investigated with care. Indeed, as discussed earlier, three conduction mechanisms generate the current and are therefore generating their own noise. The modeling of each noise and finally the total modeled noise have been reported with lines in **Figure 16**. The noise stemming from the front and back interfaces is originating from the interface traps [5], like for the inversion-mode MOSFET while the noise from the SOI layer and access resistances is explained in terms of fundamental fluctuations of the mobility of the Hooge model [3]. Contrary to the inversion-mode MOSFET, the front interface does not contribute to the total noise. The lower noise can be attributed to a change in the origin of the noise stemming from the channel, with the SOI region becoming the main contributor to the noise with regard to the channel and a significant shift toward the high gate voltage to turn the accumulation layer on. It is worth mentioning that the advantage of the accumulation-mode MOSFET is effective for high doping concentration; otherwise, the accumulation layer will act like an inversion one [36] since it implies no contribution from either the back interface or the SOI layer to the total current.

7. Conclusion

In this chapter, we reviewed several ways to suppress the low-frequency noise of MOSFETs and, in turn, the noise of analog and digital circuits. One of the most underrated approaches is to optimize the contacts and interconnects by the means of low-resistive materials, so that their contribution to the total noise can be drastically reduced. It has also been shown that a great deal must be paid to the fabrication processes. Indeed, the use of processes demonstrating very low damage generation at all the stages of fabrication can lead to MOSFETs with better performances and especially reduced noise level due to a reduction of induced defects located at the gate stack and its surroundings. Additionally, these low-defect processes based on the damage-free, low-temperature, high-density plasma technology achieved a further reduction by means of the disappearance of one component of the low-frequency noise, the induced mobility fluctuations one, bringing the noise of MOSFETs to the sole fluctuations of the insulator charges. Focusing on a different electronic structure can also achieve low-noise MOSFETs. For example, minimizing the interaction carrier-traps by moving away the channel – so that fewer traps are activated and less variations of the insulator charge are generated – can achieve a reduced noise. Unfortunately, this reduction is obtained at the cost of a degradation of the electrical performances. So, the most promising structure to suppress noise is in the form of the accumulation-mode MOSFETs. Indeed, in addition to offer reduced low-frequency noise when compared to conventional MOSFETs, their electrical performances are greatly improved. These devices obviously feature various assets, which should consequently pave the road for a new era of very low noise and high-performance MOSFETs and bring microelectronic manufacturers back to the realization of highly performances and high-speed analog and digital circuits.

Author details

Philippe Gaubert[1*] and Akinobu Teramoto[2]

*Address all correspondence to: gaubert@m.tohoku.ac.jp

1 Graduate School of Engineering, Tohoku University, Sendai, Japan

2 New Industry Creation Hatchery Center, Tohoku University, Sendai, Japan

References

[1] Gaubert P, Teramoto A, Cheng W, Ohmi T. Relation Between the Mobility, 1/f Noise, and Channel Direction in MOSFETs Fabricated on (100) and (110) Silicon-Oriented Wafers. IEEE Trans Electron Devices. 2010;57(7):1597–607.

[2] Vandamme EP, Vandamme LKJ, Claeys C, Simoen E, Schreutelkamp RJ. Impact of Silicidation on the Excess Noise Behavior of MOS Structures. Solid-State Electron. 1995;38(11):1893–97.

[3] Hooge FN. 1/f Noise Sources. IEEE Trans Electron Devices. 1994;41(11):1926–35.

[4] Mc. Whorter AL. Semiconductor Surface Physics, University of Pennsylvania Press, Philadelphia, PA. 1957:207.

[5] Mikoshiba H. 1/f Noise in n-Channel Silicon-Gate MOS Transistors. IEEE Trans Electron Devices. 1982;29(6):965–70.

[6] Ghibaudo G, Roux-dit-Buisson O, Nguyen-Duc Ch, Brini J. Improved Analysis of Low Frequency Noise in Field-Effect MOS Transistors. Solid Stat Solidi A. 1991;124:571.

[7] Kuroda R, Yonezawa A, Teramoto A, Li TL, Tochigi Y, Sugawa S. A Statistical Evaluation of Random Telegraph Noise of In-Pixel Source Follower Equivalent Surface and Buried Channel Transistors. IEEE Trans Electron Devices. 2013;60(10):3555–61

[8] Gaubert P, Teramoto A, Kuroda R, Nakao Y, Tanaka H, Sugawa S, et al. Analysis of the Low-Frequency Noise Reduction in Si(100) Metal-Oxide-Semiconductor Field-Effect Transistors. Jpn J Appl Phys. 2011;50:04DC01.

[9] Von Haartman M, Lindgren AC, Hellstrom PE, Malm BG, Zhang SL, Ostling M. 1/f Noise in Si and Si0.7Ge0.3 pMOSFETs. IEEE Trans Electron Devices. 2003;50(12):2513–9.

[10] Teramoto A, Hamada T, Yamamoto M, Gaubert P, Akahori H, Nii K, et al. Very High Carrier Mobility for High-Performance CMOS on a Si(110) Surface. IEEE Trans Electron Devices. 2007;54(6):1438–45.

[11] Le Royer C. Interfaces and Performance: What Future for Nanoscale Ge and SiGe Based CMOS? Microelectron Eng. 2011;88(7):1541–8.

[12] Deng F, Johnson RA, Asbeck PM, Lau SS, Dubbelday WB, Hsiao T, et al. Silicidation Process Using NiSi and Its Device Application. J Appl Phys.1997;81(12): 8047–51.

[13] Kuroda R, Tanaka H, Nakao Y, Teramoto A, Miyamoto N, Sugawa S, et al. Ultra-low Series Resistance $W/ErSi_2/n^+$-Si and $W/Pd_2Si/p^+$-Si S/D Electrodes for Advanced CMOS Platform. IEDM. 2010. Proc. International Electron Devices Meeting; San Francisco, USA; 26.2.1–26.2.4

[14] Isogai T, Tanaka H, Goto T, Teramoto A, Sugawa S, Ohmi T. Formation and Property of Yttrium and Yttrium Silicide Films as Low Schottky Barrier material for n-Type Silicon. Jpn J Appl Phys. 2008;47(4):3138–41.

[15] Gaubert P, Teramoto A, Hamada T, Yamamoto M, Kotani K, Ohmi T. 1/f Noise Suppression of pMOSFETs Fabricated on Si(100) and Si(110) Using an Alkali-Free Cleaning Process. IEEE Trans Electron Devices. 2006;53(4):851–6.

[16] Kumagai Y, Abe K, Fujisawa T, Watabe S, Kuroda R, Miyamoto N, et al. Large-Scale Test Circuits for High-Speed and Highly Accurate Evaluation of Variability and Noise in Metal-Oxide-Semiconductor Field-Effect Transistor Electrical Characteristics. Jpn J Appl Phys. 2011;50:106701 (Copyright © 2011 The Japan Society of Applied Physics).

[17] Ohmi T, Hirayama M, Teramoto A. New Era of Silicon Technologies Due to Radical Oxidation Based Semiconductor Manufacturing. J Phys D: Appl Phys. 2006;39(1):R1–17.

[18] Hamada T, Teramoto A, Akahori H, Nii K, Suwa T, Hirayama M, et al. High Performance Low Noise CMOS Fabricated on Flattened (110) Oriented Si Substrate. Proc. AWAD. 2004:163–166. Proc. Asia–Pacific Workshop on Fundamentals and Applications of Advanced Semiconductor Devices; Nagasaki, Japan; 163–166

[19] Watabe S, Teramoto A, Abe K, Fujisawa T, Miyamoto N, Sugawa S, et al. Statistical Evaluation of Process Damage Using an Arrayed Test Pattern in a Large Number of MOSFETs. IEEE Trans Electron Devices. 2010;57(6):1310–8.

[20] Ohmi T. Total Room Temperature Wet Cleaning for Si Substrate Surface. Journal of The Electrochemical Society 1996;143(9):2957–64.

[21] Nishiuchi K, Oka H, Nakamura T, Ishikawa H, Shinoda M (Eds). A Normally-Off Type Buried Channel MOSFET for VLSI Circuits. International Electron Devices Meeting, San Francisco, CA, USA; 1978.

[22] Nguyen TN, Plummer JD (Eds). Physical Mechanisms Responsible for Short Channel Effects in MOS Devices. International Electron Devices Meeting, Washington DC, USA; 1981.

[23] Yuan T, Hu GJ, Dennard RH, Terman LM, Chung-Yu T, Petrillo KE. A Self-Aligned 1-mm-Channel CMOS Technology with Retrograde n-Well and Thin Epitaxy. IEEE Trans Electron Devices. 1985;32(2):203–9.

[24] Suzuki H, Kuroda R, Teramoto A, Yonezawa A, Sugawa S, Ohmi T (Eds). Impact of Random Telegraph Noise Reduction with Buried Channel MOSFET. The 2011 International Conference on Solid State Devices and Materials, Nagoya, Japan; 2011.

[25] Yonezawa A, Teramoto A, Obara T, Kuroda R, Sugawa S, Ohmi T, editors. The Study of Time Constant Analysis in Random Telegraph Noise at the Sub-Threshold Voltage Region. International Reliability Physics Symposium, Monterey; 2013.

[26] Watabe S, Sugawa S, Teramoto A, Ohmi T. New Statistical Evaluation Method for the Variation of Metal-Oxide-Semiconductor Field-Effect Transistors. Jpn J Appl Phys. 2007;46(4B):2054–7.

[27] Kumagai Y, Abe K, Fujisawa T, Watabe S, Kuroda R, Miyamoto N, et al. Large-Scale Test Circuits for High-Speed and Highly Accurate Evaluation of Variability and Noise in Metal-Oxide-Semiconductor Field-Effect Transistor Electrical Characteristics. Jpn J Appl Phys. 2011;50: 106701.1–106701.11.

[28] Watabe S, Teramoto A, Abe K, Fujisawa T, Miyamoto N, Sugawa S, et al. A Simple Test Structure for Evaluating the Variability in Key Characteristics of a Large Number of MOSFETs. IEEE Trans Semicond Manuf. 2012;25(2):145–54.

[29] Abe K, Sugawa S, Watabe S, Miyamoto N, Teramoto A, Kamata Y, et al., editors. Random Telegraph Signal Statistical Analysis Using a Very Large-scale Array TEG with 1M MOSFETs. IEEE Symposium on VLSI Technology, Kyoto; 2007.

[30] Kuroda R, Yonezawa A, Teramoto A, Li TL, Tochigi Y, Sugawa S. A Statistical Evaluation of Random Telegraph Noise of In-Pixel Source Follower Equivalent Surface and Buried Channel Transistors. IEEE Trans Electron Devices. 2013;60(10):3555–61.

[31] Colinge J-P, Lederer D, Afzalian A, Yan R, lee C-W, Akhavan ND, et al. Properties of Accumulation-Mode Multi-Gate Field-Effect Transistors. Jpn J Appl Phys. 2009;48(3): 034502.

[32] Cheng W, Teramoto T, Ohmi T. Experimental Demonstration and Analysis of High Performance and Low 1/f Noise Tri-gate MOSFETs by Optimizing Device Structure. Microelectron Eng. 2009;86(6/7):1786–8.

[33] Kuroda R, Teramoto A, Cheng W, Sugawa S, Ohmi T. Hot Carrier Instability Mechanism in Accumulation-Mode Normally-off SOI n-MOSFETs and Their Reliability Advantages. ECS Trans. 2007;6(4):851–6.

[34] Akarvardar K, Cristoloveanu S, Dufrene B, Gentil P, Schrimpf R, Blalock B, et al. Evidence for Reduction of Noise and Radiation Effect in g4-FET Depletion All-Around Operation. Proc. 35th European Solid-State Device Research Conference; Grenoble, France 2005:89–92.

[35] Gaubert P, Teramoto A, Sugawa S, Ohmi T. Hole Mobility in Accumulation Mode Metal-Oxide-Semiconductor Field-Effect Transistors. Jpn J Appl Phys. 2012;51:04DC07.

[36] Gaubert P, Teramoto A, Sugawa, Ohmi T. Low Frequency Noise Assessment of Accumulation Si p-MOSFETs. Proc. SSDM. Proc. International Conference on Solid State Devices and Materials; Kyoto, Japan 2012:839–40.

Chapter 2

Symplectic Principal Component Analysis: A Noise Reduction Method for Continuous Chaotic Systems

Min Lei and Guang Meng

Additional information is available at the end of the chapter

http://dx.doi.org/10.5772/64410

Abstract

In real environments, data obtained from a dynamic system were more or less contaminated by noise. In addition, the dynamic system was often very complicated, non-linear, even chaotic, and sometimes unknown. These might generally lead the measured data to be very complex and seemingly stochastic. In order to obtain the more intrinsic characteristics of the dynamical system, especially chaotic systems, from the measured data, how to effectively reduce noise was still a crucial issue. The aim of this chapter is to introduce a method of noise reduction based on symplectic geometry for the continuous chaotic systems with noise called symplectic principal component analysis (SPCA). The symplectic geometry was a kind of phase space geometry that could preserve the dynamical structure of the system, especially non-linear structure. In symplectic space, the SPCA method could give the dominant principal component values of the data and the component values of the noise floor by the measure-preserving symplectic transform. In the end, this chapter investigated the performance of the SPCA method and applied it to reduce noise in the chaotic time series and experimental data.

Keywords: symplectic geometry, noise reduction, chaotic continuous systems, symplectic principal component analysis (SPCA), principal component analysis (PCA), locally projective non-linear noise reduction (NNR), time series, sunspot, signal processing

1. Introduction

Data measured from a system were often very complicated since the underlying dynamical system was usually complex, non-linear, stochastic, or even unknown. And it was generally heavily corrupted by noise in experimental situations or real environments so that the data might be treated as noise and disregarded. Since chaotic phenomena had been discovered, interpretation of irregular dynamics of various systems as a deterministic chaotic process had been popular and widely studied in almost all fields of science and engineering, such as health sciences, nanosciences, physical sciences, economics, ecology, biomedicine, fault diagnosis, and so on. The seemingly random data had been reanalysed. A number of important algorithms based on chaos theory had been employed to distinguish between the chaotic data and noise, or reduce noise from the data, or infer the system dynamics from the data [1–6]. However, it was still challenging how to get the most information of the underlying dynamical system from a measured data. Meanwhile, how to appropriately reduce noise from a measured data was a first crucial issue.

At present, a number of noise reduction methods had been used for noisy contaminated chaotic times series [7–10]. These methods had mainly employed the delay-embedded time series to reduce the noise of a measured data. In other words, the analysed data were firstly reconstructed into a phase space according to Takens' embedding theorem [11–12]. Then, various approaches were applied to deal with the reconstructed phase space, such as the singular value decomposition (SVD)-based denoising method. In the study field of noise reduction, the SVD method was one of the most commonly used methods to reduce noise from a time series. However, the heart of the SVD was linear in nature so that it might become misleading technique when it dealt with a non-linear time series [13]. For this, we proposed a novel method based on symplectic geometry, which was non-linear in nature.

The symplectic geometry was a kind of phase space geometry that could preserve the system structure, especially non-linear structure. Since Kang [14] had proposed a symplectic algorithm for solving symplectic differential, the symplectic geometry method had been widely used to investigate the equation solving problems of various complex dynamical systems in physics, mathematics, classical mechanics, quantum mechanics, elasticity and Hamiltonian mechanics, and so on [15–25]. For some bottleneck basic problems in elasticity, the novel symplectic approaches were explored and developed by Zhong and his group [17, 18] and references therein and Lim et al. [19–25] to study the numerical solutions of the plates and beams. Some researchers had also studied eigenvalue problems of the Hamilton matrix in symplectic geometry [26–34]. For main eigenvalues of a large Hamiltonian matrix, an inverse substation method and an adjoint symplectic inverse substation method had been proposed [26–27]. The symplectic elementary transformation had been used to solve the eigenvalues of the Hamilton matrices, because it could preserve the structures of the Hamiltonian matrices [28–33]. A new algorithm (SROSH) for computing the SR factorization was proposed by optimal symplectic Householder transformations [28]. A detailed error analysis of the (SROSH) method was described by Salam and Al-Aidarous [30]. For sparse and large-structured matrices, some modified versions were usually involved in structure-preserving Krylov subspace-type methods [33]. The SR decomposition could be obtained using a symplectic QR-like decomposition [29–32] or symplectic Gram-Schmidt algorithm [34]. More results on numerical aspects of symplectic Gram-Schmidt algorithms could be found in the study of Salam [34]. These methods based on symplectic geometry had mainly been used to solve the eigenvalue problems of the $2n \times 2n$ real matrices or Hamiltonian matrices in the systems of dynamical mechanics and control theory. To our knowledge, a few literatures had employed symplectic geometry theory to analyse the data generated from the non-linear systems [35–39]. The purpose of this chapter was to use symplectic geometry method to reduce noise from the chaotic time series and experimental data.

The remainder of this chapter was organized as follows. Section 2 introduced the symplectic principal component analysis (SPCA) method based on Householder transform. Section 3 provides the reduced noise method based on symplectic geometry and its algorithm. In Section 4, numerical and experimental data were described. Section 5 was devoted to analysing the noise reduction of these data. In Section 6, a general conclusion was given.

2. Symplectic principal component analysis (SPCA)

In the real life, the studied system could usually give one observable that was a noisy one-dimensional (1D) signal sampled with a finite precision. Little was known about a system equation or its phase portrait. One only could reconstruct the original system from the sampled noisy 1D signal to study the dynamical characteristics of the system. First, a time series was constructed into a trajectory matrix X (i.e., an attractor in the phase space) in terms of Takens' embedding theorem. Second, the reconstructed attractor was transformed into a Hamiltonian matrix in symplectic space. Then, the symplectic transformations were used to deal with the Hamiltonian matrix [40–41].

2.1. Phase space reconstruction (Attractor reconstruction)

A dynamical system was defined in phase space R^d. A discretized trajectory X at times $t = n^t{}_s$, $n = 1, 2, \dots$ might be described by maps of the form

$$\boldsymbol{x}_{n+1} = \boldsymbol{f}(\boldsymbol{x}_n) \tag{2.1}$$

where $x_1, x_2, \cdots, x_n$ is the measured data, that is, the observable of the system under study, t_s is sampling interval, n is the number of samples, and d is the dimension of phase space. According to Takens' embedding theorem, if d was large enough, then under certain genericity assumptions, a one-to-one image of the original set $\{x\}$ was given from the time series. The time-delay embedding approach could map the time series $\{x\}$ into a d-dimensional space R^d:

$$X = \begin{bmatrix} X_1^T \\ X_2^T \\ \vdots \\ X_m^T \end{bmatrix} = \begin{bmatrix} x_1 & x_2 & \cdots & x_d \\ x_2 & x_3 & \cdots & x_{d+1} \\ \vdots & \vdots & \cdots & \vdots \\ x_m & x_{m+1} & \cdots & x_n \end{bmatrix} \tag{2.2}$$

where $m = n - d + 1$ is the number of dots in d-dimensional reconstruction attractor, $X_{m \times d}$ denoted the trajectory matrix of the dynamical system in phase space, that is, the attractor in phase space. The matrix X contained all the dynamical information of the system that generated the data x.

2.2. Basic theory and concept of symplectic geometry

In common, the referred space was Euclidean space R_n. The architecture of Euclidean space was dependent on the bilinear symmetrical non-singular cross product:

If $x \neq 0$,

$$\begin{aligned} &\langle \boldsymbol{x}, \boldsymbol{x} \rangle > 0 \\ \therefore \quad &\|\boldsymbol{x}\| = \sqrt{\langle \boldsymbol{x}, \boldsymbol{x} \rangle} > 0. \end{aligned} \tag{2.3}$$

This showed that the measurement of Euclidean space was length scale. Unlike Euclidean space, the concept of orthogonal cross course existed in symplectic space. Symplectic space was the space with a special symplectic structure and dependent on a bilinear antisymmetric non-singular cross product—symplectic cross product:

$$[x, y] = \langle x, Jy \rangle, \tag{2.4}$$

where

$$J = J_{2n} = \begin{bmatrix} 0 & +I_n \\ -I_n & 0 \end{bmatrix}.$$

When $n = 1$, there was

$$[\boldsymbol{x}, \boldsymbol{y}] = \begin{vmatrix} x_1 & y_1 \\ x_2 & y_2 \end{vmatrix}.$$

In traditional meaning, odd dimension concept did not exist in the non-singular antisymmetric matrix, and hence, symplectic space was even dimension.

Therefore, the measurement of symplectic space was area scale. The length of arbitrary vector in symplectic space always equalled to zero and without any signification. This was essential difference between symplectic and Euclidean spaces [14–16, 42]. In symplectic geometry, the symplectic similar transform was the regular transform, which could preserve measure and keep the essential character of the primary time series unchanged, so it was fit to analyse non-linear system process.

Definition 2.1. *Let S be a matrix, if* $JSJ^{-1} = S^{-*}$, *then S was a symplectic matrix.*

Definition 2.2. *Let H be a matrix, if* $JHJ^{-1} = -H^{*}$, *then H was a Hamilton matrix.*

Theorem 2.1. *Any n × n matrix could be made into a Hamilton matrix. Let a matrix as* $A_{n\times n}$, *if the matrix M could be constructed as follows:*

$$M = \begin{pmatrix} A & 0 \\ 0 & -A^{*} \end{pmatrix} \tag{2.5}$$

M is a Hamilton matrix.

Proof:

$$\begin{aligned} \because \quad & J\begin{pmatrix} A & 0 \\ 0 & -A^{*} \end{pmatrix}J^{-1} \\ &= \begin{pmatrix} 0 & I_n \\ -I_n & 0 \end{pmatrix}\begin{pmatrix} A & 0 \\ 0 & -A^{*} \end{pmatrix}\begin{pmatrix} 0 & I_n \\ -I_n & 0 \end{pmatrix}^{-1} \\ &= \begin{pmatrix} -A^{*} & 0 \\ 0 & A \end{pmatrix} \\ &= -\begin{pmatrix} A & 0 \\ 0 & -A^{*} \end{pmatrix}^{*} \end{aligned}$$

∴ According to Definition 2.2, M was a Hamilton matrix.□

Theorem 2.2. *Hamilton matrix kept unchanged at symplectic similar transform.*

Proof:

Let S as a symplectic transform matrix, H as a Hamilton matrix. Then, S^{-1} is also symplectic matrix. According to Definitions 2.1 and 2.2, there was

$$\begin{aligned} & J\left(SHS^{-1}\right)J^{-1} \\ &= JSJ^{-1}JHJ^{-1}JS^{-1}J^{-1} \\ &= S^{-*}\left(-H^{*}\right)S^{*} \\ &= -\left(SHS^{-1}\right)^{*} \end{aligned}$$

∴ SHS^{-1} is also a Hamilton matrix.

∴ $SHS^{-1} \sim H$

So, Hamilton matrix H kept unchanged at symplectic similar transform.

Theorem 2.3. *Let* $M \in C^{2d \times 2d}$ *be Hamilton matrix, so* e^M *was symplectic matrix.*

Theorem 2.4. *The product of sympletcic matrices was also a symplectic matrix.*

Proof:

Let $S_1, S_2, \ldots, S_n$ as symplectic matrix, respectively. According to Definition 2.1, there were

$$
\begin{aligned}
&JS_1J^{-1} = S_1^{-*} \\
&JS_2J^{-1} = S_2^{-*} \\
&\cdots \\
&JS_nJ^{-1} = S_n^{-*}
\end{aligned}
$$

$$
\begin{aligned}
&J(S_1S_2\cdots S_n)J^{-1} \\
&= JS_1J^{-1}JS_2J^{-1}J\cdots J^{-1}JS_nJ^{-1} \\
&= S_1^{-*}S_2^{-*}\cdots S_n^{-*} \\
&= (S_1S_2\cdots S_n)^{-*}
\end{aligned}
$$

So, the product of sympletcic matrixes was also a symplectic matrix.

Theorem 2.5 *Suppose a Householder matrix H was*

$$
H = H(k,\omega) = \begin{pmatrix} P & 0 \\ 0 & P \end{pmatrix}, \tag{2.6}
$$

$$
\begin{aligned}
&P = I_n - \frac{2\varpi\varpi^*}{\varpi^*\varpi}, \\
&\varpi = (0,\cdots,0;\omega_k,\cdots,\omega_n)^T \neq 0.
\end{aligned}
$$

So, H is a symplectic unitary matrix. ϖ^* is ϖ conjugate transposition.

Proof:

In order to prove that the matrix H was symplectic matrix, we only needed to prove $H^*JH = J$.

$$
\begin{aligned}
H^*JH &= \begin{pmatrix} P & 0 \\ 0 & P \end{pmatrix}^* J \begin{pmatrix} P & 0 \\ 0 & P \end{pmatrix} \\
&= \begin{pmatrix} 0 & P^*P \\ -P^*P & 0 \end{pmatrix}
\end{aligned} \tag{2.7}
$$

$$\because P = I_n - \frac{2\varpi\varpi^*}{\varpi^*\varpi}$$

$$\therefore \quad P^* = P$$

$$\begin{aligned} P^*P &= P^2 \\ &= \left(I_n - \frac{2\varpi\varpi^*}{\varpi^*\varpi}\right)\left(I_n - \frac{2\varpi\varpi^*}{\varpi^*\varpi}\right) \\ &= I_n - \frac{4\varpi\varpi^*}{\varpi^*\varpi} + \frac{4\varpi(\varpi^*\varpi)\varpi^*}{(\varpi^*\varpi)(\varpi^*\varpi)} \\ &= I_n \end{aligned} \tag{2.8}$$

where $\boldsymbol{\varpi} = (0, \cdots, 0; \omega_k, \dots, \omega_n)^T \neq 0$

Plugging Eq. (2.8) into Eq. (2.7), we had:

$$H^*JH = J \tag{2.9}$$

$\therefore$ H is a symplectic matrix.

$$\begin{aligned} H^*H &= \begin{pmatrix} P & 0 \\ 0 & P \end{pmatrix}^* \begin{pmatrix} P & 0 \\ 0 & P \end{pmatrix} \\ &= \begin{pmatrix} P^*P & 0 \\ 0 & P^*P \end{pmatrix} \\ &= I_{2n} \end{aligned} \tag{2.10}$$

$\therefore$ H is also a unitary matrix.

$\therefore$ The Householder matrix H is a symplectic unitary matrix.

In the real calculation, the above householder matrix H could be constructed from a time series according to Theorem 2.6.

Theorem 2.6. *Suppose x and y were two unequal n dimension vectors, and* $\|x\|_2 = \|y\|_2$, *so there was elementary reflective array* $H = 1 - 2\,\varpi\varpi^T$, *which made* $Hx = y$, *where* $\omega = \frac{x - y}{\|x - y\|_2}$.

For a non-zero n dimension vector $x = (x_1, x_2, \cdots, x_n)^T$ we could note $\alpha = \|x\|_2$. Then, there was

$$Hx = \alpha e_1, \tag{2.11}$$

$$\begin{cases} H = 1 - 2\varpi\varpi^T \\ e_1 = (1, 0, \cdots, 0)^T \\ \varpi = \dfrac{1}{\rho}(x - \alpha e_1) \\ \rho = \|x - \alpha e_1\|_2 \end{cases} \tag{2.12}$$

Then $\|\omega\|_2 = 1$, and H is elementary reflective array.

It was easy to testify that the elementary reflective array H was symmetry matrix ($H^T = H$), orthogonal matrix ($H^T H = 1$) and involution matrix ($H^2 = 1$).

2.3. Building Hamiltonian matrix

For a time series x, the covariance matrix A of the matrix X (see Eq. (2.2)) could first be given as follows:

$$A = \bar{X} \cdot \bar{X}^T \tag{2.13}$$

Where $\bar{X}$ was a mean-centred matrix by removing the mean values of the columns of the data matrix A is a $d \times d$ symmetrical matrix. Then, the Hamiltonian matrix M could be constructed by Eq. (2.5).

2.4. Symplectic principal component analysis (SPCA)

For Hamilton matrix M ($2n \times 2n$), its eigenvalues could be obtained by symplectic similar transform, such as symplectic QR decomposition. At present, there had been some algorithms for symplectic QR decomposition [28–34, 43]. Here, we could use symplectic Householder transform instead of the matrix Q to decompose the matrix M:

$$M = QR \tag{2.14}$$

where Q is a Householder matrix, that is, a symplectic unitary matrix (the eigenvector matrix). R is an upper triangular matrix. It was easy to prove that the Householder matrix Q was a symplectic unitary matrix according to Theorem 2.5.

In theory, for a Hamiltonian matrix $M = \begin{pmatrix} A & 0 \\ 0 & -A^T \end{pmatrix}$ (see Eq. (2.5)), there was a Householder matrix H, which made HMH^T as an upper Hessenberg matrix, namely

$$HMH^T = \begin{pmatrix} P & 0 \\ 0 & P \end{pmatrix} \begin{pmatrix} A & 0 \\ 0 & -A^T \end{pmatrix} \begin{pmatrix} P & 0 \\ 0 & P \end{pmatrix}^T \tag{2.15}$$

$$= \begin{pmatrix} PAP^T & 0 \\ 0 & -PA^TP^T \end{pmatrix}$$

$$= \begin{pmatrix} B & 0 \\ 0 & -B^T \end{pmatrix}$$

$$\therefore \quad \lambda(A) = \lambda(B) \tag{2.16}$$

Thus, the primary $2n$-dimension space could be transformed into n-dimension space to resolve. The symplectic eigenvalues ($\boldsymbol{\mu} = \{\mu_1, \mu_2, \cdots, \mu_{2d}\}$) of the matrix M could be composed of those ($\lambda(A) = \{\mu_1, \mu_2, \cdots, \mu_d\}$) of the matrix A. The symplectic eigenvectors of the Householder matrix H could consist of those of the matrix P corresponding to the symplectic eigenvalues of the matrix A. These eigenvalues were given by descending order as following:

$$\mu_1 > \mu_2 > \cdots > \mu_l >> \mu_{l+1} \geq \cdots \geq \mu_d \tag{2.17}$$

The above symplectic eigenvalue method could be used to analyse the principal components of a system, called symplectic principal component analysis (SPCA) [40].

3. Noise reduction based on SPCA

For the above Hamiltonian matrix M, its symplectic eigenvalues $\boldsymbol{\mu} = \{\mu_1, \mu_2, \cdots, \mu_{2d}\}$ could be got by using symplectic QR decomposition in Section 2.4. According to Eq. 2.17, the dominant symplectic eigenvalues μ_i ($i = 1, \ldots, l$) of A could be obtained, that is

$$\mu_1 > \mu_2 > \cdots > \mu_l \tag{3.1}$$

when $\mu_i >> \mu_{i+1}$. The lower symplectic eigenvalues μ_i, ($i = l+1, \ldots, d$) got into a noise floor. Consequently, noise could be reduced from the data x by eliminating the lower eigenvalue components μ_i, ($i = l+1, \ldots, d$). Corresponding to the dominant components μ_i, ($i = 1, \ldots, l$), the truncation matrix W could be got from the Householder matrix H. The new data with reduced noise could be generated using W to transform the original data x. The procedure of the SPCA method was given as follows:

1. Reconstruct the trajectory matrix X from the raw data x in terms of Eq. (2.2);
2. Build the real $d \times d$ symmetry matrix A (see Eq. (2.13));
3. Calculate the matrix P of the Householder matrix H from the matrix A [37]. Let A as follows:

$$A = \begin{pmatrix} a_{11} & a_{12} & \cdots & a_{1n} \\ a_{21} & a_{22} & \cdots & a_{2n} \\ \vdots & \vdots & \cdots & \vdots \\ a_{n1} & a_{n2} & \cdots & a_{nn} \end{pmatrix} = \begin{pmatrix} a_{11} & A_{12}^{(1)} \\ \alpha_{21}^{(1)} & A_{22}^{(1)} \end{pmatrix} \tag{3.2}$$

First, suppose $\alpha_{21}^{(1)} \neq 0$, otherwise this column would be skipped and the next column would be considered until the ith column of $\alpha_{2i}^{(1)} \neq 0$. Set first column vector of A:

$$S^{(1)} = \left(a_{11}^{(1)}, a_{21}^{(1)}, \cdots, a_{n1}^{(1)}\right)^T = \left(a_{11}, a_{21}, \cdots, a_{n1}\right) \tag{3.3}$$

Then, the elementary reflective matrix $P^{(1)}$ is computed by:

$$P^{(1)} = I - 2\varpi^{(1)}(\varpi^{(1)})^T \tag{3.4}$$

where

$$\begin{cases} \alpha_1 = \left\|S^{(1)}\right\|_2 \\ E^{(1)} = (1, 0, \cdots, 0)^T \qquad n \times 1 \\ \rho_1 = \left\|S^{(1)} - \alpha_1 E^{(1)}\right\| \\ \varpi^{(1)} = \dfrac{1}{\rho_1}\left(S^{(1)} - \alpha_1 E^{(1)}\right) \end{cases} \tag{3.5}$$

So, A is changed to a matrix $A^{(2)}$:

$$P^{(1)}A = \begin{pmatrix} \sigma_1 & a_{12}^{(2)} & \cdots & a_{1n}^{(2)} \\ 0 & a_{22}^{(2)} & \cdots & a_{2n}^{(2)} \\ \vdots & \vdots & \ddots & \vdots \\ 0 & a_{n2}^{(2)} & \cdots & a_{nn}^{(2)} \end{pmatrix} = A^{(2)} \tag{3.6}$$

where the first element of the first column is σ_1. Other elements were zeros.

Then, the second column vector of $A^{(2)}$ is also given like the above way. Let

$$S^{(2)} = \left(0, a_{22}^{(2)}, \cdots, a_{n2}^{(2)}\right)^T \tag{3.7}$$

construct $P^{(2)}$ matrix:

$$P^{(2)} = I - 2\omega^{(2)}(\varpi^{(2)})^T \tag{3.8}$$

where

$$\begin{cases} \alpha_2 = \left\|S^{(2)}\right\|_2 \\ E^{(2)} = (0, 1, 0, \cdots, 0)^T \qquad n \times 1 \\ \rho_2 = \left\|S^{(2)} - \alpha_2 E^{(2)}\right\| \\ \varpi^{(2)} = \dfrac{1}{\rho_2}\left(S^{(2)} - \alpha_2 E^{(2)}\right) \end{cases} \tag{3.9}$$

Thus, the second column of $A^{(2)}$ is also changed to all zero vector except the first and second elements. $A^{(3)}$ is obtained:

$$P^{(2)}A^{(2)} = A^{(3)} \tag{3.10}$$

By repeating above-mentioned method, the matrix P could be obtained until $A^{(n)}$ became an upper triangle matrix:

$$P = P^{(n)}P^{(n-1)}\cdots P^{(1)} \tag{3.11}$$

It was easy to show that P was a symmetrical, orthogonal and involution matrix.

4. Use the matrix P to transform the matrix A into the upper triangular matrix T. The absolute values of the diagonal components T_{ii} by descending order were called as the symplectic eigenvalues of the matrix A (see Eq. (2.17));

5. Corresponding to the dominant symplectic principal component values μ_i, (i= 1,..., l), let the first l column vectors of W as those of P. Corresponding to the lower eigenvalues μ_i, (i=l+1,...,d), the other vectors of W were zeros;
6. Construct the symplectic transform matrix Q and the Hamiltonian matrix S, i.e.

$$Q = \begin{pmatrix} W & 0 \\ 0 & W \end{pmatrix} \tag{3.12}$$

$$S = \begin{pmatrix} X & 0 \\ 0 & -X \end{pmatrix} \tag{3.13}$$

where X is given by Eq. (2.2).

7. Use the Q project S into Y,

$$Y = Q^T S \tag{3.14}$$

8. Reestimate the $\hat{X}$ with reduced-noise,

$$\hat{X} = QY \tag{3.15}$$

The reduced-noise data could be given by the first row of the matrix $\hat{X}$.

For the heavily noisy time series, the first estimation of data was usually not good. Here, one could repeat the above steps 7 and 8 several times. Generally, the second or third reconstructed data would be better than the first reconstructed data.

4. Numerical and experimental data

This chapter employed the three typical chaotic equations [41].

1. Lorenz equation

$$\begin{cases} \dot{x} = \sigma(y - x) \\ \dot{y} = \gamma x - y - xz \\ \dot{z} = -bz + xy \end{cases} \tag{4.1}$$

where $\sigma = 10$, $b = 8/3$, and $\gamma = 28$.

2. Duffing equation

$$\ddot{x} + c\dot{x} - \varepsilon x(1 - x^2) = A\cos(\omega t) \tag{4.2}$$

where ε=1, c=0.4, A=0.4, and ω=1.

3. Chua's equation

$$\begin{aligned} \dot{x} &= \alpha(y - x - f(x) \\ \dot{y} &= x - y + z \\ \dot{z} &= -\beta y \\ f(x) &= bx + \frac{a-b}{2}(|x+1| - |x-1|) \end{aligned} \tag{4.3}$$

where α = 15.6, β = 28.58, a = -8/7, and b = -5/7.

A measurement noise e was used because all the real measurements were polluted by noise. Here, the Gaussian white noise e as a measure noise was added to the clean signal x generated from the above chaotic equations. The contaminated data x_n is obtained as follows:

$$x_n(t) = x(t) + e(t) \tag{4.4}$$

The signal-to-noise ratio was defined by the following:

$$SNR = 10\log(\sigma_x^2 / \sigma_n^2) \tag{4.5}$$

where σ_x and σ_n are the standard deviation of the clean signal x and the noise e, respectively. The more details of noise notions were referred to the literature [44–46]. Here, SNR is 10 dB.

As for a practical example of noise reduction, we chose the sunspot number data obtained from monthly observations. Sunspot number series were short, highly non-linear and noisy [49]. It was hard to predict accurately the sunspot period, although Wolf had reported the well-known 11-year cycle.

5. Noise reduction analysis based on SPCA

5.1. Performance evaluation of SPCA

SPCA, like PCA, could not only represent the original data by capturing the relationship between the variables but also reduce the contribution of errors in the original data. Here, the performance evaluation of SPCA was first given based on the analysis of Lorenz chaotic time series [39].

5.1.1. Performance evaluation on the representation analysis

Since the real systems were usually unknown, it was necessary to study the influence of sampling time, data length, and noise on the representation analysis based on the SPCA approach. For a clean chaotic time series, the root mean square error (RMSE) as a measure was employed in order to estimate the difference between the clean original data and the SPCA-filtered data:

$$RMSE = \sqrt{\frac{1}{N}\sum_{i=1}^{N}[x(i) - \hat{x}(i)]^2} \tag{5.1}$$

where $x(i)$ and $\hat{x}(i)$ are the clean original data and the estimated data, respectively.

In order to analyse noisy data, the percentage of principal components (PCs) was defined to study the occupancy rate of each PC in order to reduce noise as follows:

$$P_i = \frac{\mu_i}{\sum_{i=1}^{d}\mu_i} \times 100\% \tag{5.2}$$

where d is the embedding dimension and μ_i is the ith principal component value.

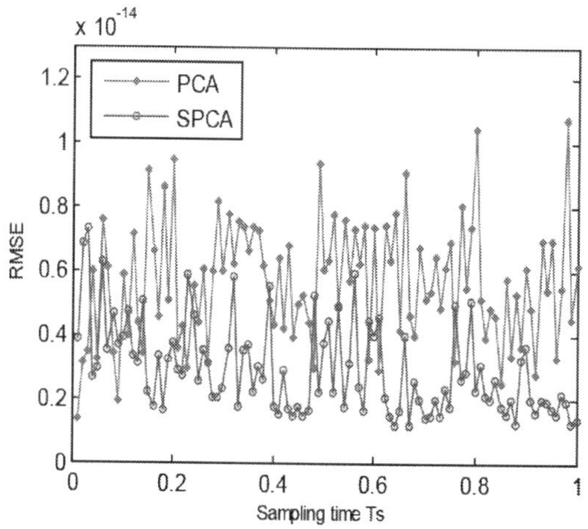

Figure 1. RMSE versus sampling time curves for the SPCA and PCA.

Here, we took the Lorenz system as an example. The dimension of the Lorenz system was 3, then, the embedding dimension d of the matrix A was chosen as 8. For the clean Lorenz time series generated by Eq. (4.1) (i.e., no noise $e = 0$ in Eq. (4.4)), when $k = d$, the estimated data were obtained by SPCA and PCA with $k=d$, respectively. For the different sampling time T_s, the RMSE values are calculated in **Figure 1** by Eq. (5.1). The RMSE values of SPCA were lower than 10^{-14} for the different sampling time (see **Figure 1**). The results showed that the SPCA method was better than the PCA. **Figure 2** shows the RMSE values of the different data lengths for SPCA and PCA, respectively. For the different data lengths, the RMSE values of SPCA were also lower than 10^{-14} (see **Figure 2**). From the **Figures 1** and **2**, we could see that the sampling time and the data length had less effect on SPCA method in the case of free noise.

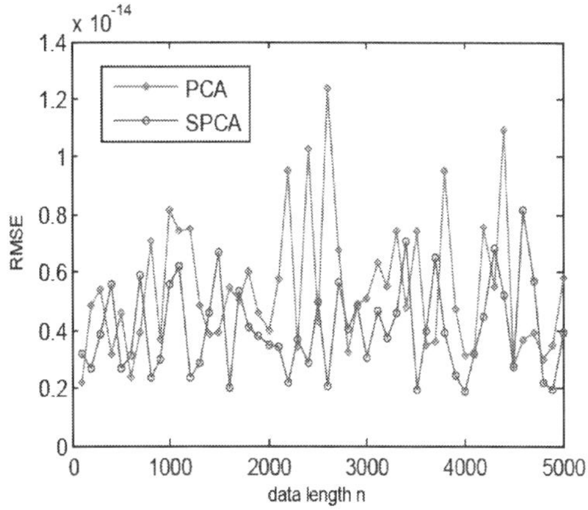

Figure 2. RMSE versus data length curves for the SPCA and PCA.

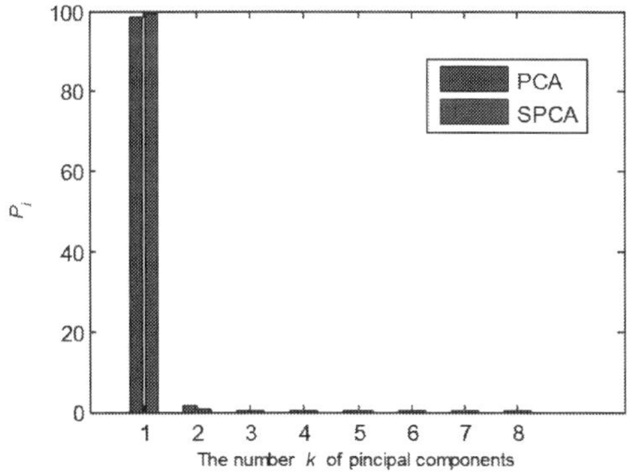

Figure 3. The percentage of PCs for the SPCA and PCA.

From **Figure 3**, all of the principal components were given by Equation. (5.2) for the clean Lorenz time series according to the SPCA and PCA methods, respectively. We could see that the first largest symplectic principal component (SPC) of the SPCA was a little larger than that of the PCA. For the SPCA method, the first largest SPC was almost possessed of all the proportion of the SPCs. Next, the reduced space spanned by a few largest SPCs was explored to estimate the chaotic Lorenz time series without noise. For the different data length, we gave the RMSE values between the original data and the data estimated from the first seven largest SPCs and PCs, respectively, that is, in the case of $k = 7$ (see **Figure 4**). The sampling time is 0.1. The result showed that the data length had less effect on the SPCA than the PCA. **Figure 5** shows the effect of sampling time on different number of PCs for the SPCA and the PCA methods, respectively. When the PCs number k =1 and k =7, respectively, the change of RMSE values with the sampling time is given in **Figure 5**. We could see that the RMSE values of the SPCA were smaller than those of the PCA. The sampling time had also less impact on the SPCA than the PCA.

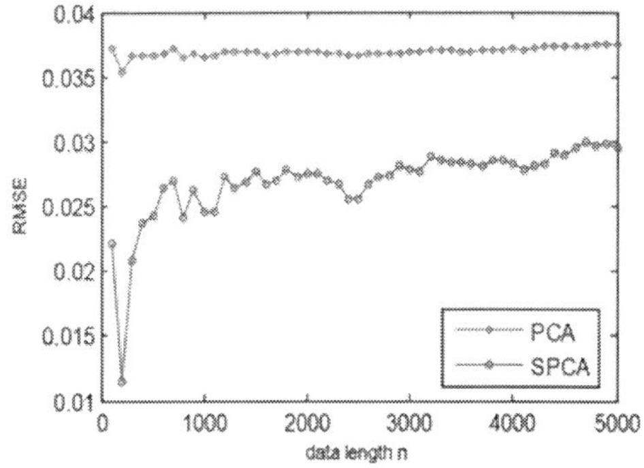

Figure 4. The RMSE versus the data length for the SPCA and PCA, where k =7. The sampling time is 0.1.

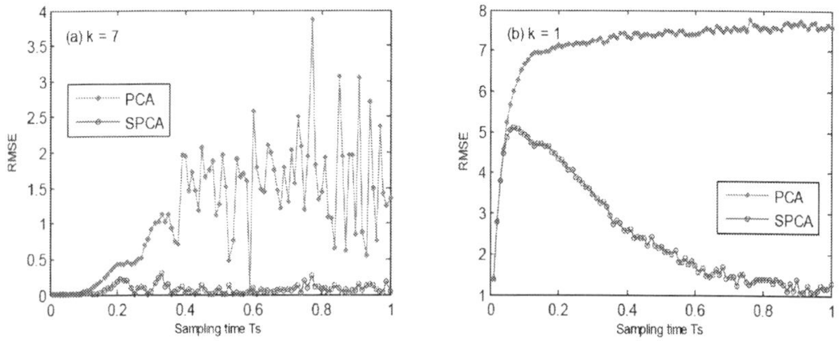

Figure 5. The RMSE values versus the sampling time for the SPCA and PCA, where (a) the PCs number k =7; (b) k =1.

Furthermore, the estimated data based on the first three largest SPCs are calculated in **Figure 6**, where the original data x are given with a sampling time of 0.01 from chaotic Lorenz system. The average error between the original data and the estimated data was −6.55e-16. The corresponding standard deviation was 1.03e-2. The estimated data were very close to the original data not only in time domain (see **Figure 6a**) but also in phase space (see **Figure 6b**).

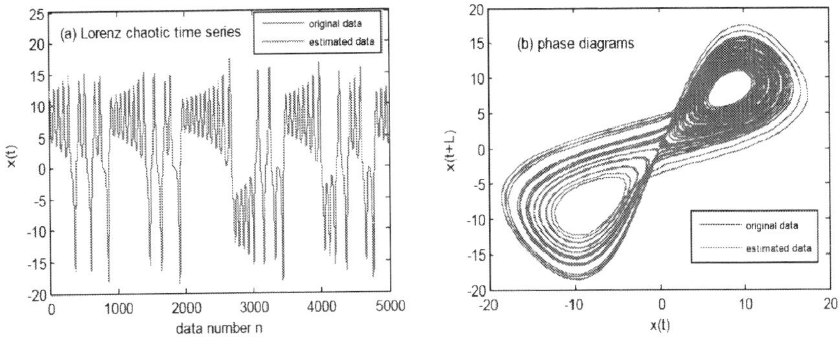

Figure 6. Chaotic signal reconstructed by the proposed SPCA algorithm with k=3, where (a) the time series of the original Lorenz data x without noise and the estimated data; (b) phase diagrams with L = 11 for the original Lorenz data x without noise and the estimated data. The sampling time T_s = 0.01.

To sum up, we could see that the SPCA method preserved the essential dynamical character of the primary time series generated from chaotic continuous systems. These indicated that the SPCA could reflect intrinsic non-linear characteristics of the original time series. The SPCA could elucidate the dominant features in the observed data. It was feasible for the SPCA to study the principal component analysis (PCA) of time series. Moreover, the SPCA would help to retrieve dominant patterns from the noisy data.

5.1.2. Performance evaluation on noise reduction

For the performance evaluation on noise reduction, the SPCA method was further studied by dealing with the noisy Lorenz data x with Gaussian white noise of zero mean and one variance. The phase diagrams of the noisy and clean data are given in **Figure 7a** and **b**. The clean data x were obtained by the sampling time 0.01 from the chaotic Lorenz system with noise-free. The noisy data were the chaotic Lorenz data x with Gaussian white noise of zero mean and one variance (Eqs. (4.1) and (4.4)). The time delay L is 11 in **Figure 7**. It was obvious that noise is very strong (see **Figure 7a**). Here, we first built an attractor X with the embedding dimension of 8. Then, the transform matrix W was constructed when k=1. The first denoised data are generated according to Section 3 (see **Figure 7c** and **d**). In **Figure 7c**, the first denoised data are compared with the noisy Lorenz data x from the view of time field. **Figure 7d** shows the corresponding phase diagram of the first denoised data. Compared with **Figure 7a**, the first denoised data could basically give the structure of the original system. In order to obtain better results, these denoised data were reduced noise again by the step (8). For the second noise reduction, the results were greatly improved in **Figure 7e** and **f**. Comparing by **Figure 7c**, **d**, **e** and **f**, the curves of the second denoised data were better than those of the first denoised data whether in time domain or in phase space. **Figure 7g** shows that the PCA technique gave the first denoised result. Like **Figure 7e**, the second denoised data are obtained by the PCA (see **Figure 7h**). Although some of noise had been further reduced in **Figure 7h**, the curve of PCA was not better than that of SPCA in **Figure 7e**. The reason was that the PCA was a linear method indeed. When non-linear structures had to be considered, it could be misleading, especially in the case of a large sampling time (see **Figure 8**). The used program code of the PCA came from the TISEAN tools (http://www.mpipks– dresden.mpg.de/~tisean).

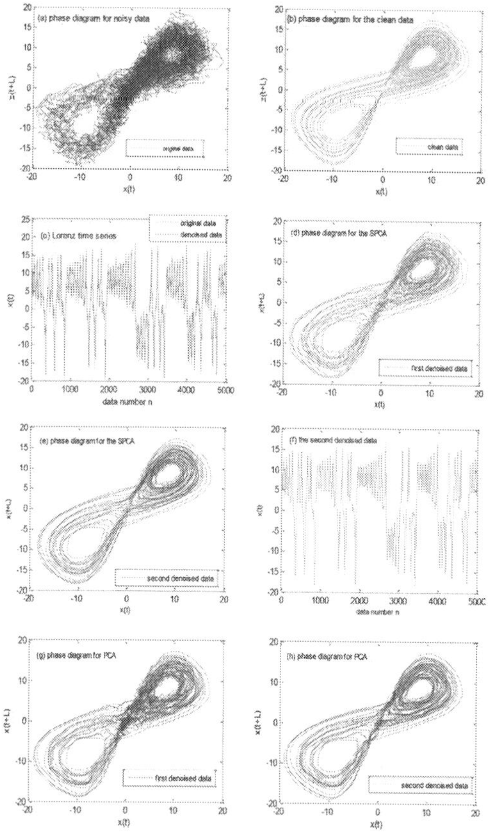

Figure 7. The noise reduction analysis of the proposed SPCA algorithm and PCA for the noisy Lorenz time series, where L=11. (a) Phase diagram for noisy data, (b) phase diagram for the clean data, (c) Lorenz time series, (d) phase diagram for the SPCA, (e) phase diagram for the SPCA, (f) the second denoised data, (g) phase diagram of the first denoised data for PCA and (h) phase diagram of the second denoised data for PCA.

In order to evaluate the effectiveness of noise reduction, the correlation dimension D_2 was estimated by the Grassberger-Procaccia's algorithm [47–48] because it manifested non-linear properties of chaotic systems. The variations of correlation dimension D_2 with embedding dimension d were given for the clean, noisy, and denoised Lorenz data (see **Figure 8**). The sampling time was 0.1. The results showed that, for the clean and SPCA denoised data, the trend of the curves had a platform and tended to smooth in the vicinity of 2. For the noisy data, the trend of the curve was constantly increasing and had no platform. For the PCA denoised data, the trend of the curve was also increasing and trended to a platform with 2. However, this platform was smaller than that of SPCA. The PCA algorithm was less effective than the SPCA algorithm. The result indicated that it was difficult for the PCA to describe the non-linear structure of a system.

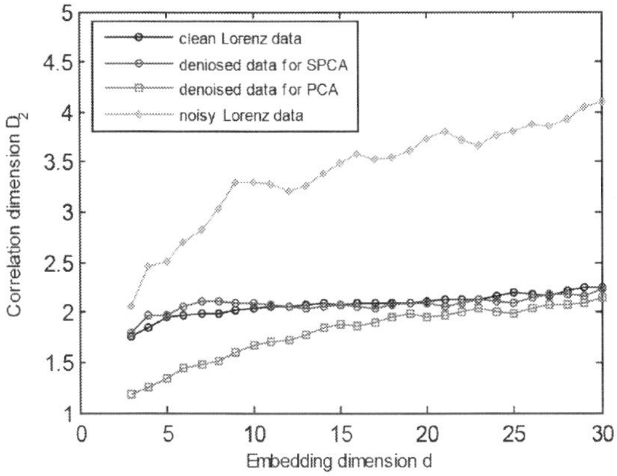

Figure 8. D_2 versus embedding dimension d.

Besides, it was necessary to show that the method of the locally projective non-linear noise reduction (NNR) in the TISEAN package (http://www.mpipks-dresden.mpg.de/~tisean) could not give the better result than SPCA and PCA [41].

5.2. Noise reduction based on SPCA

In the noise level SNR = 10 dB, the noisy Duffing chaotic data (see **Figure 9a**) and the noisy Chua's chaotic data (see **Figure 9c**) show reduced noise by applying the SPCA. Embedding dimension d=8 had been used with the time delay k=1. The reduced noise results of SPCA are shown in **Figure 9**. The third denoised data of the noisy Duffing chaotic data are shown in **Figure 9b**. For the noisy Chua's chaotic data, the third denoised data are given in **Figure 9d**. Obviously, a lot of noise had been removed from the noisy time series. Here, the number of dominant components was chosen as one according to the curves of SPCs in **Figures 10** and **11**.

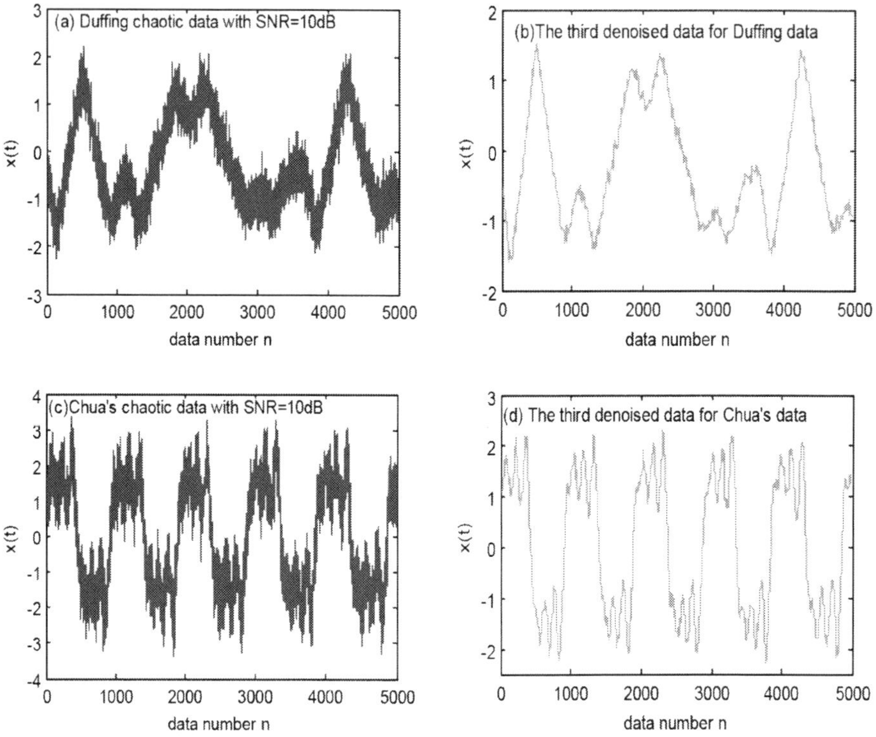

Figure 9. The noise reduction for the noisy Duffing chaotic data and the noisy Chua's chaotic data based on the SPCA. (a) Duffing chaotic data with SNR=10 dB, (b)The third denoised data for Duffing data, (c)Chua's chaotic data with SNR=10 dB and (d) The third denoised data for Chua's data.

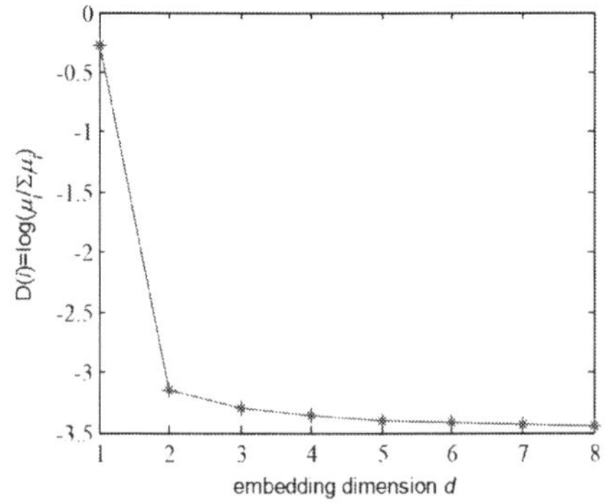

Figure 10. The symplectic principal component analysis of the noisy Duffing chaotic data x.

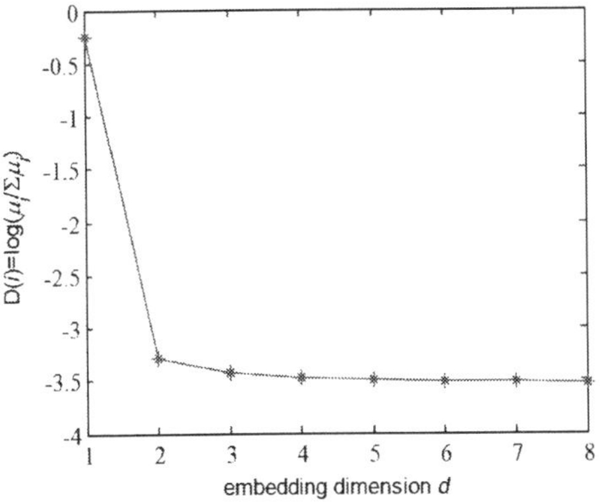

Figure 11. The symplectic principal component analysis of the noisy Chua's chaotic data x.

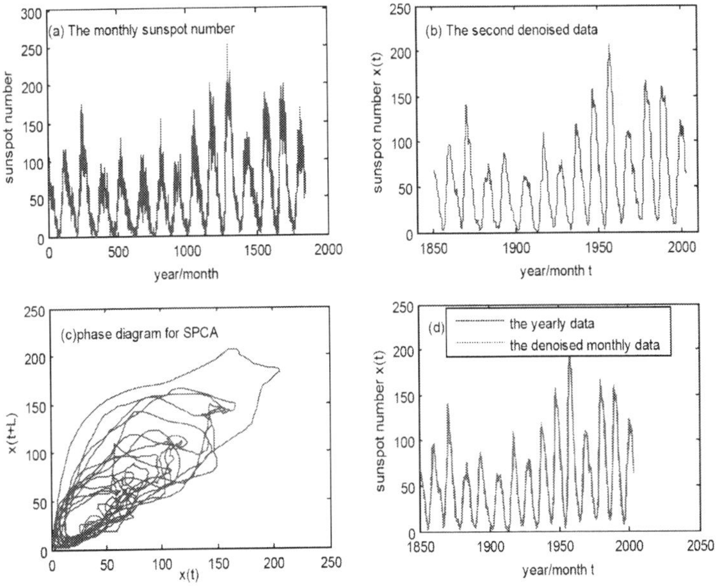

Figure 12. The noise reduction analysis of the monthly sunspot number based on the SPCA. (a) The monthly sunspot number, (b) the second denoised data, (c) phase diagram for SPCA, (d) the yearly sunspot data and the denoised monthly sunspot data.

The SPCA method was also applied to reduce noise from the monthly sunspot data (see **Figure 12a**). The time range was from January 1850 to February 2004. There was a lot of noise in the monthly data. According to our SPCA noise reduction algorithm, the resultant data were given by reducing noise twice when d =8, k=1 (see **Figure 12b**). It could be seen that plenty of noise had been removed from the monthly sunspot numbers. Its attractor in phase space was clearly shown with L=12 in **Figure 12c**. It was obvious that the sunspot cycle could be explained neither by regular periodicity nor by a sequence of random process. The sunspot numbers contained non-linear characteristics [49]. Furthermore, we compared the denoised monthly data and the yearly data in **Figure 12d**. The denoised monthly data were very close to the yearly data. The results showed that SPCA could effectively remove the noise from the monthly sunspot data.

Here, the first symplectic principal component was chosen to reduce the noise in the monthly sunspot numbers x referring to **Figure 13**.

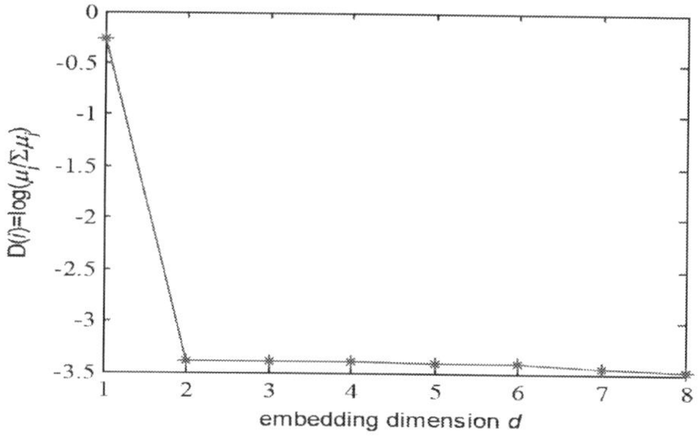

Figure 13. he symplectic principal component analysis of the monthly sunspot numbers x.

6. Conclusion

This chapter had proposed the symplectic principal component analysis method (SPCA) and the noise reduction method based on SPCA. In theory, the SPCA method was intrinsically non-linear, which could reflect non-linear structure of non-linear dynamical systems very well. Therefore, the clean chaotic Lorenz data were used to study the performance of the SPCA method by calculating RMSE, percentage, correlation dimension, and phase space diagrams. The results showed that the SPCA method could yield more reliable results for chaotic time series under the different data lengths and sampling times, especially with short data length and undersampled sampling time, than the classic PCA. With regard to noise reduction, SPCA algorithm was also more effective than PCA and NNR. It could reduce more noise than the other two methods. And for the SPCA noise reduction, the denoised data could catch the non-linear structure of the systems. The SPCA method was used to remove noise from the noisy Lorenz data, Duffing data, Chua's data, and sunspot data. The results showed that the SPCA algorithm had a good effect of noise reduction. It was suitable for the SPCA method to analyse the non-linear time series and deal with the noisy data.

Acknowledgements

This work was supported by the Science Fund for Creative Research Groups of the National Natural Science Foundation of China (Grant No. 51421092), the National Natural Science Foundation of China (Grant No. 10872125), Research Fund of State Key Laboratory of Mechanical System and Vibration (Grant No. MSV-MS-2010-08), Research Fund from Shanghai Jiao Tong University for medical and engineering science. (Grant No. YG2013MS74), Key Laboratory of Hand Reconstruction, Ministry of Health, Shanghai, People's Republic of China, Shanghai Key Laboratory of Peripheral Nerve and Microsurgery, Shanghai, People's Republic of China.

Author details

Min Lei* and Guang Meng

*Address all correspondence to: leimin@sjtu.edu.cn

Institute of Vibration, Shock and Noise, State Key Laboratory of Mechanical System and Vibration, Fundamental Science on Vibration, Shock and Noise Laboratory, Shanghai Jiao Tong University, Shanghai, P. R. China

References

[1] Gao, J. B., Cao, Y. H., Tunge, W. -W., and Hu, J., Multiscale analysis of complex time series-Integration of chaos and random fractal theory and beyond, New York, Wiley Interscience, 2007.

[2] Kantz, H., and Schreiber, T., Nonlinear time series analysis, Cambridge, 2nd edition, 2003, ISBN: 0521529026.

[3] Hu, J., Gao, J. B., and Tung, W. W., Characterizing heart rate variability by scale-dependent Lyapunov exponent, Chaos, vol. 19, Article ID 028506, 2009.

[4] Shin, K., Hammond, J. K., and White, P. R., Iterative SVD method for noise reduction of low-dimensional chaotic time series. Mechanical Systems and Signal Processing, vol. 13, no. 1, pp. 115–124, 1999.

[5] Ozer, G. and Ertokatli, C., Chaotic processes of common stock index returns: an empirical examination on Istanbul Stock Exchange (ISE) market. African Journal of Business Management, vol. 4, no. 6, pp. 1140–1148, 2010.

[6] Bogris, A., Argyris, A., and Syvridis, D., Encryption efficiency analysis of chaotic communication systems based on photonic integrated chaotic circuits. IEEE Journal of Quantum Electronics, vol. 46, no. 10, pp. 1421–1429, 2010.

[7] Gao, J.B., Sultan, H., Hu, J., and Tung, W.W., Denoising nonlinear time series by adaptive filtering and wavelet shrinkage: a comparison. IEEE Signal Processing Letters, vol. 17, pp. 237–240, 2010.

[8] Orzesko, W., The new method of measuring the effects of noise reduction in chaotic data. Chaos, Solitons and Fractals, vol. 38, pp. 1355–1368, 2008.

[9] Wang, W. J., and Chen, J., Estimation and application of correlation dimension of experimental time series. Journal of Vibration and Control, vol. 7, pp. 1035–1047, 2001.

[10] Sun, J., Zhao, Y., Zhang, J., Luo, X., and Small, M., Reducing colored noise for chaotic time series in the local phase space. Physical Review E, vol. 76, 026211, 2007.

[11] Hegger, R., Kantz, H., and Schreiber, T., Practical implementation of nonlinear time series methods: the TISEAN package. Chaos, vol. 9, no. 2, pp. 413–435, 1999.

[12] Takens, F., Detecting Strange Attractors in Turbulence. In: Dynamical Systems and Turbulence, eds. D. A. Rand and L. -S. Young, Lecture Notes in Mathematics, vol. 898 (Springer-Verlag, New York) pp. 366–381, 1980.

[13] Palus, M. and Dvorak, I., Singular-value decomposition in attractor reconstruction: pitfalls and precautions. Physica D, vol. 55, pp. 221–234, 1992.

[14] Kang, F. On difference schemes and symplectic geometry. In: Proceeding of the 1984 Beijing Symposium on Differential Geometry and Differential Equations—COMPUTATION OF PARTIAL DIFFERENTIAL EQUATIONS, ed. Feng Kang (Science Press, Beijing), pp. 42–58, 1985.

[15] Kang, F., Symplectic geometry and numerical methods in fluid dynamics. in: Proceedings of the 10th International Conference on Numerical Methods in Fluid Dynamics, Beijing, China, June 23–27, 1986, in F. G. Zhuang, Y. L. Zhu(Eds), Lecture Notes in Physics, vol.264, Springer Berlin Heidelberg, 1986, pp1-7, ISBN: 978-3-540-17172-0 (Print) 978-3-540-47233-9 (Online).

[16] Kang, F. and Qin, M.Z., The symplectic methods for the computation of Hamiltonian equations. In: Proceedings of 1st Chinese Conference on Numerical Methods of PDE's, March 1986, Shanghai, eds. Zhu You-lan, Guo Ben-yu, Lecture Notes in Mathematics, vol. 1297, Springer, Berlin, 1997, p. 1.

[17] Zhong, W., Plane elasticity problem in strip domain and Hamiltonian system. Journal of Dalian University of Technology, vol. 31(4), pp. 373–384, 1991 (in Chinese).

[18] Yao, W., Zhong, W. and Lim, C.W., Symplectic Elasticity. World Scientific Publishing Co. Pte. Ltd., 2009, ISBN-13 978-981-277-870-3, ISBN-10 981-277-870-5.

[19] Lim, C.W. and Xu, X.S., Symplectic elasticity: theory and applications. ASME Applied Mechanics Reviews, vol. 63(5), 050802, 2010.

[20] Lim, C.W., Cui, S., and Yao, W., On new symplectic elasticity approach for exact bending solutions of rectangular thin plates with two opposite sides simply supported. International Journal Solids Structure, vol. 44(16), pp. 5396–5411, 2007.

[21] Lim, C.W., Symplectic elasticity approach for free vibration of rectangular plates. Advances in Vibration Engineering, vol. 9(2), pp. 159–163, 2010.

[22] LÜ, C. F., Lim, C. W., and Yao, W. A., A new analytic symplectic elasticity approach for beams resting on Pasternak elastic foundations. Journal of Mechanics of Materials and Structures, vol. 4(10), pp. 1741–1754, 2009.

[23] Lim, C.W., Lü, C.F., Xiang, Y., and Yao, W., On new symplectic elasticity approach for exact free vibration solutions of rectangular Kirchhoff plates. International Journal of Engineering Science, vol. 47, pp. 131–140, 2009.

[24] Lim, C.W., Yao, W.A., and Cui, S., Benchmark symplectic solutions for bending of corner-supported rectangular thin plates. The IES Journal Part A: Civil & Structural Engineering, vol. 1(2), pp. 106 – 115, 2008.

[25] Lim, C.W. and Yao W.A., Closure on discussion of 'Benchmark symplectic solutions for bending of corner-supported rectangular thin plates' by M. Batista. The IES Journal Part A: Civil & Structural Engineering, vol. 3(1), pp. 71–73, 2010.

[26] Zhong, W. X., The inverse iteration method for the eigenproblem of large symplectic matrices. Computational Structural Mechanics and Applications, vol. 9(3), pp. 227–238, 1992.

[27] Zhong, W. X. and Zhong, X. X., On the adjoint symplectic inverse substitution method for main eigensolutions of a large Hamiltonian matrix. Journal Systems Engineering, vol. 1, pp. 41–50, 1991.

[28] Salam, A., Al-Aidarous, E., Farouk, A. EI, Optimal symplectic Householder transformations for SR decomposition. Linear Algebra and its Applications, vol. 429, pp. 1334–1353, 2008.

[29] Salam, A., Farouk, A. EI, and Al-Aidarous, E., Symplectic householder transformations for a QR-like decomposition, a geometric and algebraic approaches. Journal of Computational and Applied Mathematics, vol. 214, pp. 533–548, 2008.

[30] Salam, A. and Al-Aidarous, E., Error analysis and computational aspects of SR factorization via optimal symplectic householder transformations. Electronic Transactions on Numerical Analysis, vol. 33, pp. 189–206, 2009.

[31] Loan, C. Van, A symplectic method for approximating all the eigenvalues of a Hamiltonian matrix. Linear Algebra and its Applications, vol. 61, pp. 233–251, 1984.

[32] Bunse-gerstner, A. and Mehrmann, V., A symplectic QR-like algorithm for the solution of the real algebraic Riccati equation. IEEE Transactions on Automatic Control, vol. AC-31, pp. 1104–1113, 1986.

[33] Benner, P. and Faßbender, H., An implicitly restarted symplectic Lanczos method for the Hamiltonian eigenvalue problem. Linear Algebra and its Applications, vol. 263, pp. 75–111, 1997.

[34] Salam, A., On theoretical and numerical aspects of symplectic Gram-Schmidt-like algorithms. Numerical Algorithms, vol. 39, pp. 237–242, 2005.

[35] Xie, H.B. and Dokos, S., A hybrid symplectic principal component analysis and central tendency measure method for detection of determinism in noisy time series with application to mechanomyography. Chaos, vol. 23, p. 023131, 2013.

[36] Lei, M., Wang, Z.H., and Feng, Z.J., The application of symplectic geometry on nonlinear dynamics analysis of the experimental data. In: 14th International Conference on Digital Signal Processing Proceeding, vols. 1–2, pp. 1137–1140, 2002.

[37] Lei, M., Wang, Z.H., and Feng, Z.J., A method of embedding dimension estimation based on symplectic geometry. Physics Letters A, vol. 303(2–3), pp. 179–189, 2002.

[38] Niu, X., Qu, F., and Wang, N., Evaluating sprinters' surface EMG signals based on EMD and symplectic geometry. Journal of Ocean University of Qingdao, vol. 35(1), pp. 125–129, 2005 (in Chinese).

[39] Xie, H., Wang, Z., and Huang, H., Identification determinism in time series based on symplectic geometry spectra. Physics Letters A, vol. 342(1–2), pp. 156–161, 2005.

[40] Lei, M. and Meng, G., Symplectic principal component analysis: a new method for time series analysis. Mathematical Problems in Engineering, vol. 2011, Article ID 793429, 14pages, 2011.

[41] Lei, M., Meng, G., A Noise Reduction Method for Continuous Chaotic Systems based on Symplectic Geometry. Journal of Vibration Engineering & Technologies, vol. 3, no. 1, pp13–24, 2015.

[42] Kang, F. Difference schemes for Hamiltonian formalism and symplectic geometry. JCM, vol. 4(3), pp. 276–289, 1986.

[43] Schreiber, R. and Loan, C.V., A storage-efficient WY representation for products of Householder transformations. SIAM Journal on Scientific and Statistical Computing, vol. 10(1), pp. 53–57, 1989.

[44] Argyris J. and Andreadis I., The influence of noise on the correlation dimension of chaotic attractors. Chaos, Solitons & Fractals, vol. 9(3), pp. 343–61, 1998.

[45] Argyris J. and Andreadis I., On the influence of noise on the largest Lyapunov exponent and on the geometric structure of attractors. Chaos, Solitons & Fractals, vol. 9(6), pp. 947–58, 1998.

[46] Tung W.W., Gao J.B., Hu J., and Yang L., Recovering chaotic signals in heavy noise environments. Physical Review E, vol. 83, Article ID 046210, 2011.

[47] Grassberger P. and Procaccia I., Characterization of strange attractors. Physical Review Letters, vol. 50, p. 346, 1983.

[48] Grassberger P. and Procaccia I., Measuring the strangeness of strange attractors. Physica D, vol. 9(1–2), pp. 189–208, 1983.

[49] Lei, M. and Meng, G., Detecting nonlinearity of sunspot number. International Journal of Nonlinear Sciences and Numerical Simulation, vol. 5(4), pp. 321–326, 2004.

Chapter 3

Optimal Placement of Sensors and Actuators for Feedforward Noise and Vibration Control

Thomas Haase

Additional information is available at the end of the chapter

http://dx.doi.org/10.5772/64409

Abstract

Active systems for noise and vibration control have been under investigation for more than three decades now. It is well-known that the placement of both sensor and actuator is of crucial importance for the achievement of a reasonable noise and vibration reduction. By measuring a reference signal, a feedforward control system is able to reduce even broadband stochastic excitations. Using a simple vibration control example, this chapter summarises and analyses the parameters that influence the noise and vibration reduction of a feedforward control system. Furthermore, an optimisation tool for the placement of actuators and sensors of a feedforward control system is introduced. A special emphasis is placed on the analysis of the impacts of causality and the number of filter weights on the actuator placement. It can be shown that the actuator placement for a feedforward control system is dependent on the delays, the filter weights and even on the structural damping of the system. Besides the simulation results, the dependency of the actuator placement on the filter weights is experimentally investigated on a simple aluminium plate. The simulation results could only be partially validated at this stage due to model uncertainties.

Keywords: optimisation, feedforward control, noise and vibration reduction, causality, signal processing

1. Introduction

The research into the control of noise and vibration has been conducted for several decades now. A major concern of many research studies is the placement of sensors and actuators to optimise the system's performance [1–3]. A lot of metrics for sensor and actuator placement have been studied. Mainly, two approaches can be classified: First, the controllability- and observability-driven sensor and actuator optimisation, which is mainly based on the eigenvectors and eigenfrequencies of the structure [4, 6], and, secondly, the development of an optimisation tool that couples an optimiser and a detailed controller simulation during the optimisation [7, 8].

According to the most recent literature, the active system used to calculate the cost function of the optimisation can be either a feedforward or a feedback control system [16, 31]. With regard to feedback control, several optimisation chains can be found in which the actuator and sensor positions are optimised simultaneously with several controller parameters (e.g. weighting matrices, feedback gain). Regarding feedforward controllers, only very simple control laws are used to calculate the cost function, for example, linear quadratic optimal control theory [9–11]. However, this simple control law does not incorporate the limitations of an experimentally realisable feedforward controller into the optimisation. This would be important because aspects such as causality, finite impulse response (FIR), filter length and saturation can limit the performance of a feedforward control system.

That is why this chapter first summarises the limitations of a feedforward controller, and, secondly, explains the relevant parameters for the design of such controllers. Furthermore, this study investigates the optimisation of sensor and actuator positions for feedforward controllers. By reviewing the limitation parameters of a feedforward control system and knowing that the actuator and sensor placement has an influence on all these parameters, the following question arises: How much influence does each of these parameters have on the actuator and sensor placement? To answer this question, the optimisation tool presented here allows the optimisation of sensor and actuator positions with respect to causality and filter length. Through the variation of the causality conditions and the filter weights during the optimisation of an actuator position, the relevance of these parameters for an active vibration control (AVC) system can be shown.

This chapter is structured as follows: First of all, the experimental test bed and the simulation model are presented to give an impression of the investigated system. Secondly, the parameters that are likely to influence the feedforward controller performance are presented and analysed in detail. The next section introduces the optimisation tool and shows how optimised sensor and actuator positions can be derived specifically for feedforward control systems. The last major section discusses the simulation and experimental results. Finally, a conclusion and an outlook for future research are presented.

2. Simulation model and experimental test bed

The simulations and experiments are conducted on a simple aluminium plate mounted on its four corners (with an offset of 35 mm from the corners in x- and y-direction). An aluminium plate seems to be a very simple system, but can actually be used to easily reproduce the presented results. The system model is described by the eigenvectors and eigenfrequencies of the plate calculated with a finite element (FE) simulation in ANSYS® with an undamped modal analysis. The structural properties used in the FE simulation are presented in **Table 1**.

http://dx.doi.org/10.5772/64409

Parameter	Property (Unit)	Parameter	Property (Unit)
Young's modulus	70 (GPa)	Element size	0.01 (m)
Poisson's ratio	0.34 (–)	Density	2700 (kg/m^3)

Table 1. Material properties of the FE simulation.

The damping properties of the model are defined by modal damping ratios. For the simulations presented here, a light damping ratio of 0.5% and a medium damping of 2.5% are used. A good description of the derivation of the transfer functions with the calculated eigenfrequencies, eigenmodes and damping ratios can be found in textbook [32]. The plate is not a complex structure like, for example, an aircraft fuselage, but is only used to generate the different transfer functions that can be easily reproduced. Indeed, by using a simple aluminium plate with different damping conditions, a variety of transfer functions can be evaluated. The size of (800 × 600 × 3) mm^3 enables a high modal density of up to 38 eigenfrequencies in the investigated frequency range from 1 up to 600 Hz. From a system theory point of view, the light damping configuration is a classical resonant system, for example, a generator housing or a fuselage panel. In contrast, the medium damping configuration is more comparable to the acoustic room responses presented, for example, by Kuo and Morgan [13]. Therefore, it can be stated that the presented study can be transferred to a lot of practical technical systems. Furthermore, these different configurations are investigated because the literature shows that the damping of a structure itself seems to be of importance for a feedforward controller [15].

The actuators used in this study are piezoelectric patch actuators from PI ceramics (DuraAct® P876.A12 [33]), which are widely used in active control systems. They are modelled with modal-based correction routines in such a way that the mass and the stiffness of the actuators are integrated into the structural model. Detailed information about the actuator coupling is given in [24].

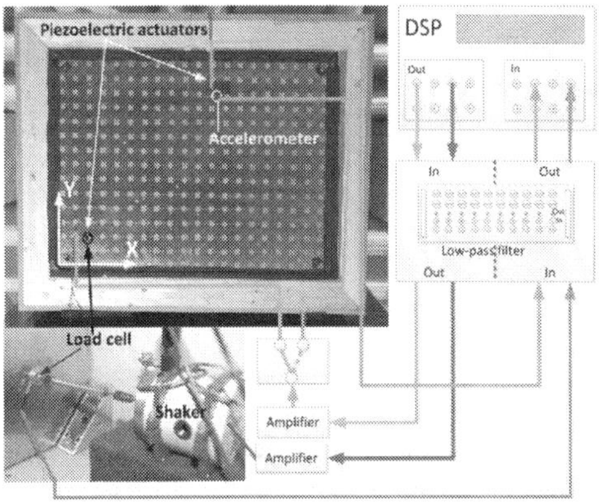

Figure 1. Experimental test bed.

For the experiments, a wooden mock-up is built to clamp the plate at its four corners. For the realisation of the medium damping, the aluminium plate is damped with constraint layer damping (CLD) from 3M®. The CLD is applied to the aluminium plate's entire rear side. A picture of the experimental test bed is given in **Figure 1**.

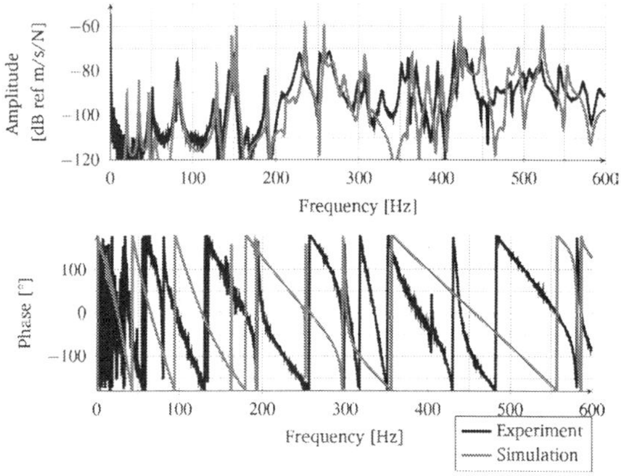

Figure 2. Bode diagram of the lightly damped secondary path.

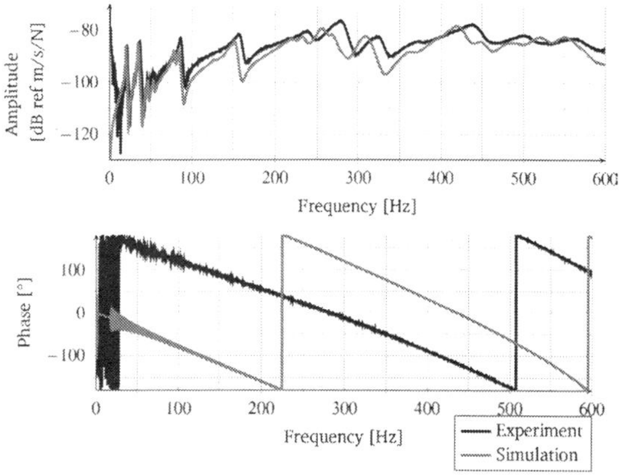

Figure 3. Bode diagram of the medium damped secondary path.

To validate the simulation model, **Figures 2** and **3** show the simulated and measured transfer paths of a piezoelectric patch actuator to an accelerometer for both damping conditions. It has to be mentioned that the phase plots of the bode diagrams only show the all-pass part of the secondary paths. The all-pass part is relevant to show the delay induced by the secondary path, while the minimum-phase part describes the phase jumps of the resonant system. The derivation of the all-pass part and the minimum-phase part of a transfer function is described in detail in [19].

Figure 2 shows that the eigenfrequencies of the lightly damped model and the curve shape are well approximated, at least the damping ratio of the simulation seems to be a bit too low. The phase response is also well approximated, regardless of the phase offsets which are in deed not relevant. The gradient of the phase response is more important because it describes the group delay of the secondary path. Above 350 Hz there is a small deviation of the gradient. The simulation model of the medium damping configuration fits the experimental model up to 300 Hz, which can be seen in **Figure 3**. Above 300 Hz, there are some deviations in the transfer path. These are caused by the simplified modelling with modal damping ratios. Furthermore, the influence of the mass and stiffness of the CLD is completely neglected during the simulation.

There are mainly two reasons why the modelled transfer paths are not modelled in more detail. The first reason is that the presented deviations are usual if the system optimisation is done in a pre-design stage and, second, for more precise structural models, a model updating or an experimentally identified structural model are necessary. Furthermore, the delays and the number of filter weights needed for modelling the structure are not very sensitive towards the accurate modal frequencies.

3. Feedforward control systems

Feedforward control systems are widely used in research [3, 12, 16] and industrial projects [17, 18]. The following subsections provide a brief introduction into the theory of feedforward controllers as well as a presentation of the calculation of an optimal causal feedforward controller [14, 19]. Afterwards, the fundamental parameters influencing the performance of a feedforward controller are described. The section closes with the presentation of the results of a huge parametric study conducted on different control set ups.

3.1. Fundamentals of feedforward control

The most common realisation of a feedforward controller is the adaptive filtered-x least mean square algorithm. The advantages of an adaptive feedforward controller are their stability (90° phase margin) and their robustness (adaptive algorithm) [13]. The analysis and description of an adaptive feedforward controller is excluded here because textbooks are available for this topic [13, 19, 25].

For stationary and ergodic excitations, the adaptive feedforward controller converges to the optimal Wiener solution [19]. Therefore, the optimal feedforward controller is used in this chapter because it is not possible to wait for the convergence of an adaptive algorithm during an optimisation.

A typical block diagram of a feedforward control system is presented in **Figure 4**. The classical feedforward controller is used to cancel a disturbance signal $d(n)$ by filtering a reference signal $x(n)$ with a filter W and the secondary path G. The secondary path G represents the dynamics of the technical system used to counteract the disturbance signal through the secondary signal $y(n)$. In the technical system presented in **Figure 1**, the secondary path is described by the reconstruction low-pass filter, the piezoelectric patch actuator, the aluminium plate, the accelerometer and the anti-aliasing filter. The primary path P describes the dynamics between the reference signal and the disturbance signal and is typically unknown in technical applications. The advantage of a feedforward controller over a feedback controller lies in the possibility to measure a time-advanced reference signal, which allows counteracting even stochastic broadband disturbances. Time-advanced means that the travelling disturbance wave (acoustic or structural bending wave) is first measured by the reference signal and after a specific delay by the disturbance or error sensor.

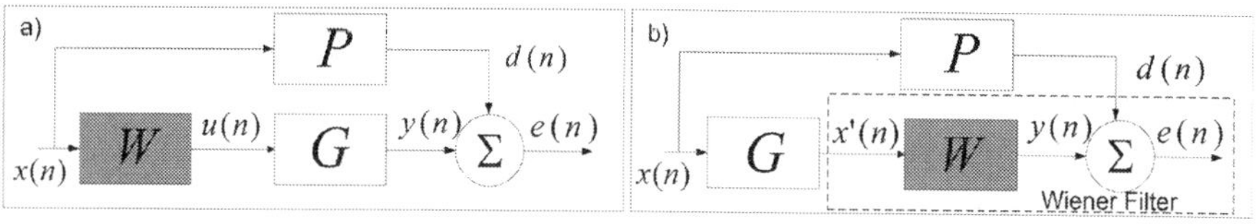

Figure 4. Block diagram of a feedforward controller: (a) classical scheme, (b) filtered reference signal scheme.

3.2. Optimal causal feedforward controller

The derivation of a causal feedforward controller is described in detail in [19]. So, only a brief summary for a single-input single-output (SISO) system is given here.

Assuming that the secondary path G and the FIR realisation of the filter W are linear and time-invariant, the chronology can be reversed as presented in **Figure 4(b)**. As can easily be seen, the classical Wiener filter theory can now be used to calculate the optimal filter for a quadratic cost function of the error signals

$$J = E[e^2(n)]. \tag{1}$$

The only step that has to be done is the calculation of the filtered reference signal $x'(n)$. The calculation of the filtered reference signal is explained by Eq. (2) in which the secondary path G is realised by a FIR filter. Of course, this is only one possibility as the secondary path can also be realised by a state-space model

$$x'(n) = \sum_{i=0}^{L-1} g_i \cdot x(n-i) \tag{2}$$

In Eq. (2), g_i describes the i-th bin of the secondary path FIR filter of length L. Referring to the Wiener filter theory [19], the optimal controller can be calculated by

$$\boldsymbol{w}_{opt} = \frac{\boldsymbol{r}_{x'd}}{\boldsymbol{R}_{x'x'}}, \tag{3}$$

in which $r_{x'd}$ is the cross-correlation vector of the filtered reference signal and the disturbance signal and $R_{x'x'}$ is the autocorrelation matrix of the filtered reference signal. The calculations of both the autocorrelation matrix and the cross-correlation vector are causal as long as only past samples are used to calculate them. The autocorrelation matrix is defined by

$$\boldsymbol{R}_{x'x'} = \begin{bmatrix} r_{x'x'}(0) & r_{x'x'}(1) & \cdots & r_{x'x'}(I-1) \\ r_{x'x'}(1) & r_{x'x'}(0) & \cdots & r_{x'x'}(I-2) \\ \vdots & & & \vdots \\ r_{x'x'}(I-1) & & \ddots & r_{x'x'}(0) \end{bmatrix}, \tag{4}$$

where I is the FIR-filter length and $r_{x'x'}(m)$is the symmetric autocorrelation function of $x'(n)$ that is defined by

$$r_{x'x'}(m) = E[x'(n)x(n-m)]. \tag{5}$$

The cross-correlation vector of $x'(n)$ and $d(n)$ is defined by

$$\boldsymbol{r}_{x'd} = [r_{x'd}(0) \quad r_{x'd}(1) \quad \cdots \quad r_{x'd}(I-1)], \tag{6}$$

in which $r_{x'd}(m)$ is defined for real and stationary time sequences by

$$r_{x'd}(m) = E[x'(n-m)d(n)]. \tag{7}$$

In summary, for the calculation of the optimal causal filter w_{opt}, it is necessary to measure or simulate:

- a secondary path model and
- synchronous time series of the reference signal $x(n)$ and the disturbance signal $d(n)$.

With the mentioned data, the filter can be calculated offline and then be implemented into the digital control system or used in the optimisation routine.

3.3. Parameters of a feedforward controller

As long as it is possible to measure a time-advanced reference signal, a feedforward control system is able to achieve a broadband error signal reduction in the presence of a broadband stationary random disturbance [13]. If the time-advance of the reference signal is larger than the delays of the secondary path, then the signal processing system is called causal. Therefore, the causality constraint describes if the time-advanced reference signal is sufficient to establish a causal or a non-causal system [13]:

$$\tau_P \geq \tau_G + \tau_\Delta \tag{8}$$

Eq. (8) is defined by the primary delay τ_P, the secondary path delay τ_G, and the signal processing delay τ_Δ. While a causal system fulfils Eq. (1), a non-causal system does not fulfil the equation. In the case of a non-causal system realisation, the performance of a feedforward controller is limited. So, to accurately simulate the performance of a feedforward controller, a causal control law, as described in Section 3.2, is necessary.

The performance of a feedforward controller is also determined by the representation of the filter *W*. Typical representations are finite impulse response filters and infinite impulse response (IIR) filters. Details about both filter representations can be found in [19] and [20]. Here, the FIR filter representation is used because of its inherent stability and also for its easy applicability for adaptive controllers.

In summary, a lot of research papers investigate feedforward control systems regarding the effects of:

- causality [15, 21, 22],
- damping [15],
- saturation [19],
- filter weights [15],
- the number of sensors and actuators [23] and
- coherence.

The references mentioned here study several effects of a feedforward control system separately. Regarding coherence, this study neglects limitations induced by a limited coherence because a high coherence is a requirement for the use of a feedforward controller. In this study, it is assumed that the coherence is approximately 1 during the experiments.

To understand all the mentioned parameters and their influence on the performance of a feedforward control system, a large parametric study has been conducted for the AVC system presented in Section 2. To the author's best knowledge, no parametric study has been published so far, which covers such a huge range of parameter variations.

3.4. Analysis methods for feedforward control systems

This section presents a small set of metrics for the analysis of feedforward control systems. These metrics are used to interpret both the parametric study and the optimised actuator positions.

First of all, a simple metric for the performance of a feedforward control system has to be defined. For reasons of simplicity, the reduction of the vibration error signal is used. Therefore, the power of the error signal in the uncontrolled state is subtracted from the controlled state

$$L_{\Delta Vib} = 10 \cdot log_{10}(\Phi_{Tot}^{On}) - 10 \cdot log_{10}(\Phi_{Tot}^{Off}) = 10 \cdot log_{10}\left(\frac{\Phi_{Tot}^{On}}{\Phi_{Off}^{tot}}\right), \quad (9)$$

where Φ_{Tot} is the frequency-weighted sum of the power spectrum of the error signal

$$\Phi_{Tot} = \Delta F \cdot \sum_{n=1}^{N_L} \Phi_{ee}(n \cdot \Delta F). \quad (10)$$

The power spectrum $\Phi_{ee}(n \cdot \Delta F)$ is calculated through the "Welch method" (Welch's averaged periodogram). In the case of systems with more than one error signal, the separate error signal spectra are averaged so that the result can be used for the calculation of the total reduction.

To analyse the length of the FIR filter and to decide whether it is sufficient to model the impulse response, the filter quality is introduced. The filter quality is defined by

$$F_Q = \frac{\sum_{i=1}^{I} |w_{opt}(n)|}{\sum_{i=1}^{I_{Ref}} |w_{opt}(n)|} \cdot 100. \quad (11)$$

The filter quality is a ratio between a reference filter of sufficient length and a shortened version of the reference filter. If the optimal filter has a short impulse response the values of F_Q are large for short filter length and if the optimal filter has a long impulse response large values of F_Q are only reached for large filter lengths.

Furthermore, as presented in Subsection 3.3, the analysis of Eq. (8) requires the derivation of the delays. This can be done by analysing the all-pass part of the measured or of the simulated transfer functions [21]. Thus, in this study the group delay [19] of the all-pass part is analysed, which can be approximated by the difference quotient and is defined by

$$\tau(j\omega) = \frac{d\varphi(j\omega)}{d\omega} \approx \frac{\varphi(j\omega_N) - \varphi(j\omega_{N-1})}{\omega_N - \omega_{N-1}}. \quad (12)$$

3.5. Parametric study

3.5.1. *Ranges of the investigated parameters*

For the parametric study, three sensors, one piezoelectric patch actuator and one disturbance force were placed on the aluminium plate to study different sensor, actuator and disturbance configurations. At this stage of the study, the placement of the elements is unimportant because the causality is varied with synthetic delays.

To study the effects of causality, damping, filter weights and saturation on the system performance, a simple system is studied first. The positions of the shaker, the actuators and, the sensors are shown in (**Figure 5**). The aluminium plate is excited by the single point force F1, which is also taken as the reference signal, and the error signal is measured at the single accelerometer S1. The secondary signal is induced via the piezoelectric patch actuator A1. The sampling frequency of the investigated system is 1200 Hz, so that the investigated bandwidth is from 1 to 600 Hz. **Table 2** presents the variations of the investigated parameters.

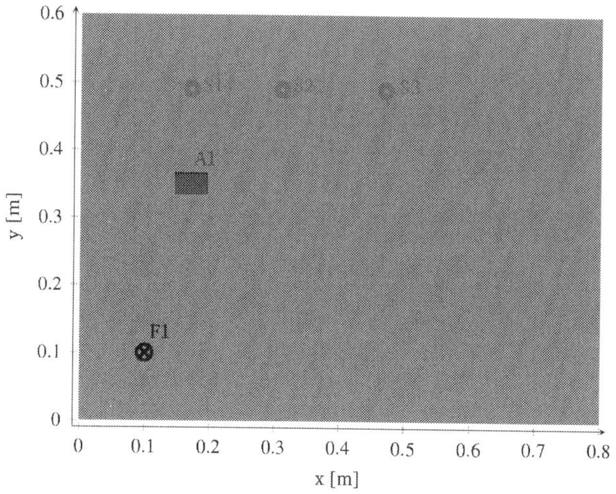

Figure 5. Configuration of sensors, actuator and disturbance source for the parametric study.

Property	**Values**
Unit sample delays β	0, ±1, ±2, ±3, ±6, ±12, ±18, ±24, ±36, ±48, ±72, ±96, ±192, ±384, ±1000, ±1300
Filter weights I	50, 100, 200, 300, 450, 600, 900, 1200, 1800, 2400, 3000, 3600, 4800
Modal damping ratio	0.5%, 2.5%
Saturation	100 V, 500 V, unlimited

Table 2. Parameter values.

To investigate different causality properties of the system, the primary or secondary path is additionally delayed by a number (β) of unit samples. In **Table 2**, a positive sign means that the primary path is delayed, meaning that the reference signal is more time-advanced. A negative sign represents a delay in the secondary path, meaning that the reference signal has a time penalty.

3.5.2. Results of the parametric study

In **Figure 6**, the vibration reduction $L_{\Delta Vib}$ is presented on the *z*-axis. The number of unit delays and the filter weights are presented on the *x*- and *y*-axis. At this stage, no saturation of the actuator voltage is introduced, which means that an unlimited voltage could be applied to the piezoelectric actuator. Furthermore, the actuator is assumed to be linear even for very high amplitudes.

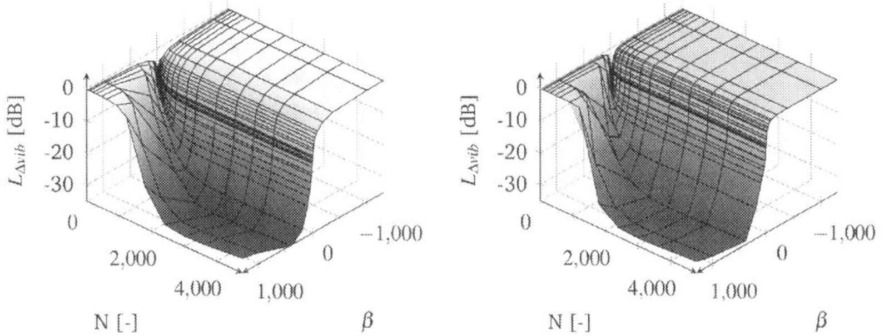

Figure 6. Total vibration reduction of the control systems with low damping (left) and medium damping (right) and no saturation.

It can be seen that in both damping configurations there is a large dependency of the vibration reduction performance on the filter weights and the delays. First of all, the lightly damped case on the left is analysed in more detail. It can be seen that even in the case of a large negative delay β, a small vibration reduction can be achieved. This is due to the lightly damped resonances that filter the stochastic force excitation towards a more deterministic signal dominated by the resonance frequencies. It can also be seen that the performance is saturated at a certain β (ca. 350), and at a large number of filter weights (ca. 1600). It should also be noticed that for small delay variations the performance is saturated at a smaller number of filter weights.

In the case of medium damping, the observed tendencies are almost identical. A major difference can be seen for negative delays, where nearly no performance is achieved. This is due to the damped resonances, which now filter the stochastic disturbances less effectively. In contrast to [15], an increased structural damping does not lead to a limitation of the performance if the feedforward controller fulfils the causality constraint.

If the actuator voltages are saturated, the overall performance is limited despite the use of a large positive delay or a large number of filter weights. **Figure 7** presents the performances for two different voltage limitations. A voltage limitation has the effect of a large performance cut-off at a certain level. For the medium damping configuration, the limitation is even more important because of the larger actuator amplitudes that are necessary to compensate the dissipated energy reasoned by the higher structural damping. If it is necessary to saturate the actuator voltage due to some design specifications (e.g. weight, size or cost), the feedforward controller, as a consequence, can be designed with shorter filters or smaller time-advances. This information is very important in designing and selecting actuators and digital signal processing (DSP) systems for feedforward control systems to reduce the overall cost and weight of such systems.

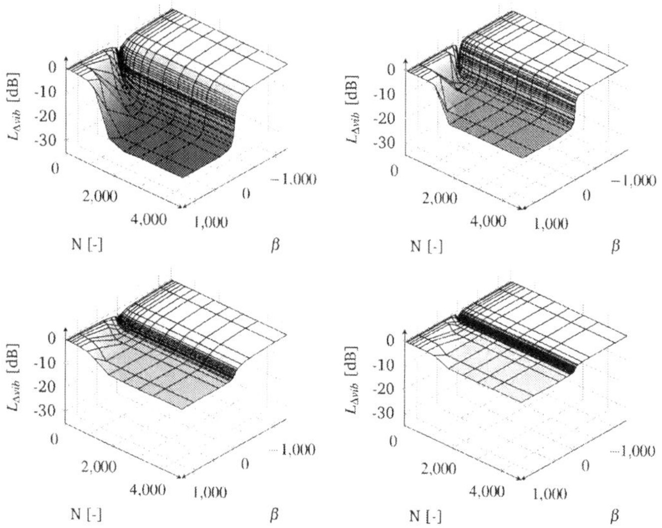

Figure 7. Total vibration reduction of the control systems with low damping (left) or medium damping (right) and a 500 V (top line) or 100 V (bottom line) limitations.

The integration of an actuator limitation is adverse, but the use of a rectangular system (more sensors than actuators) is even more dramatic for the controller performance. If the control system has to control two or three sensor positions, the performance is very limited, as presented in **Figure 8** for the lightly damped system. The introduction of a second sensor limits the performance on a scale comparable to a voltage limitation of 100 V.

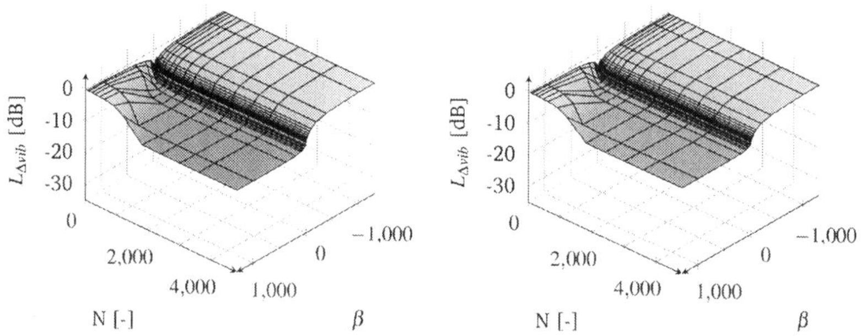

Figure 8. Total vibration reduction of the lightly damped control system with sensors S1 and S2 (left) and with sensors S1, S2 and S3 (right).

The same behaviour can be observed for the system with medium damping. The performance reduction of rectangular systems is also described in detail in [23].

If we now consider the actuator and sensor placement, it is easy to understand that the actuator and sensor placement determines the delays, the saturation (via controllability of the structure) and also the impulse responses of the primary and secondary path. Yet, by implication it is obvious that the feedforward controller parameters have an impact on the actuator and sensor placement. Therefore, the mentioned parameters are integrated in the actuator and sensor placement.

4. Sensor and actuator placement

The placement of sensors and actuators is important for the achievement of a sufficient performance of a feedforward control system. It is even more important than the type of actuator or sensor used in an active system [29]. For this reason, during the last decades numerous works on the optimisation of actuator and sensor placement have been published, e.g. see [1, 2] and various textbooks [19, 32]. Apart from the actuator and sensor placement including controllability and observability criteria, many researchers focus on a system design that combines genetic optimisation and a control law. These studies show that by using this particular system design method, the performance of a noise and vibration controller can be further improved. As a consequence, this chapter presents an optimisation routine that uses a genetic algorithm and a causal feedforward controller to optimise sensor and actuator locations. The following example of an actuator optimisation shows the influence of the parameters discussed in Section 3 on the actuator placement.

4.1. Optimisation tool

To efficiently design a feedforward system, an optimisation needs to be coupled to a realistic controller simulation. So the German Aerospace Center (DLR) developed an optimisation chain, which is presented in **Figure 9** as a flow chart.

Beside the initialisation of the excitation and the structural model, the two most important elements are the genetic optimisation routine and the calculation of the fitness function with a causal controller. These elements have to be initialised with several parameters, for example, number of sensors and actuators, the cost function and the parameters of the genetic algorithms. Therefore, a brief description of the genetic algorithm is given here. The genetic optimisation is of stochastic nature and based on Darwin's principle 'survival of the fittest'. A random population of individuals (e.g. actuator and sensor positions) is evaluated with a fitness function. The fitness function can be the local or global vibration reduction or the sound power reduction. In this study, the local vibration reduction is considered. This cost function is of relevance if the vibration at a specific location should be reduced, for example, at a mounting point to reduce vibration transmission [28]. Another example for a local reduction is a classical active noise control (ANC) system where the sound pressure at a microphone is reduced. After all individuals have been evaluated by the cost function they are ranked, selected, recombined and mutated to set up a new population, which again is evaluated. The genetic algorithms for ranking, selection, recombination and mutation are inspired by nature and meant to improve each new population compared to the old one. In this study, the genetic optimisation was aborted after 50 populations and every optimisation was performed three times to avoid singularities.

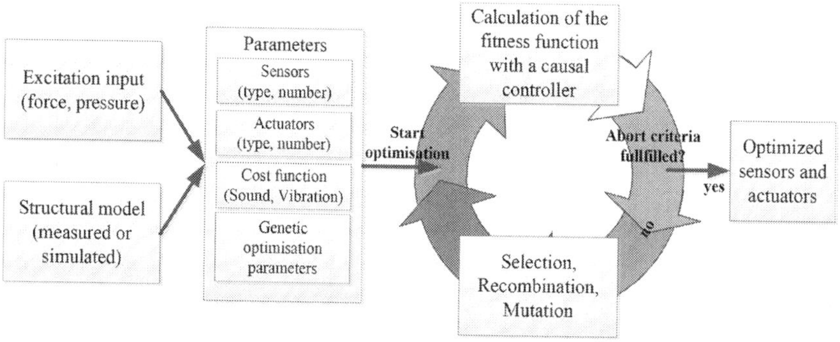

Figure 9. Flow chart of the optimization tool.

Property	Value	Property	Value
No. populations	50	Selection rate	80%
No. individuals	120	Ranking pressure	2
Mutation rate	20%	Selection mode	Stochastic universal sampling
Insertion rate	50%	Recombination	Linear

Table 3. Parameters of the genetic algorithms.

The genetic optimisation parameters are summarised in **Table 3**. Further, details about the genetic optimisation routines can be found in the manual of the genetic algorithm toolbox (GEA toolbox) [27] and in [26].

The second important element of the optimisation chain is the introduction of a causal feedforward controller. Therefore, all limitations (causality, filter weights and saturation) are integrated into the optimisation of the sensor and actuator positions. Furthermore, the delays induced by the analogue and digital signal processing units are also integrated into the simulation, as described in [21]. All signal processing elements which are relevant for the investigated system are shown in **Figure 1**. In summary, these are the anti-aliasing and reconstruction filters, the DSP, the sensors and the amplifiers. As presented in [21], the filter delays and the DSP delay are the most important, while the delay of the amplifiers and sensors is negligible. To summarise, the simulation of the feedforward controller is harmonised with the experimental realisation of a feedforward controller. The derivation of the causal controller has already been described in Section 3.2.

4.2. Actuator optimisation

In the following optimisation, the actuators are optimised to reduce the flexural vibration at a fixed error sensor position (S3 in **Figure 5**). The error sensor is fixed to facilitate the analysis of actuator position. The investigation focuses on the influence of causality, filter weights and the structural damping. To study the influence of causality, two different control systems are investigated:

- A non-causal system (NCS): The feedforward controller is designed with a simplified secondary path G*, where only the structural dynamics of the aluminium plate without any analogue and digital signal processing delays are integrated. For the controller design, the reference signal $x(n)$ and a delayed version of the disturbance signal $d(n)$. are used. The additional delay of the signal is 20 samples, which is sufficient to realise causal systems all over the plate for the chosen error sensor position.
- A causal system (CS): The feedforward controller is designed with the secondary path G that includes all filters, etc., the reference signal $x(n)$ and the disturbance signal $d(n)$.

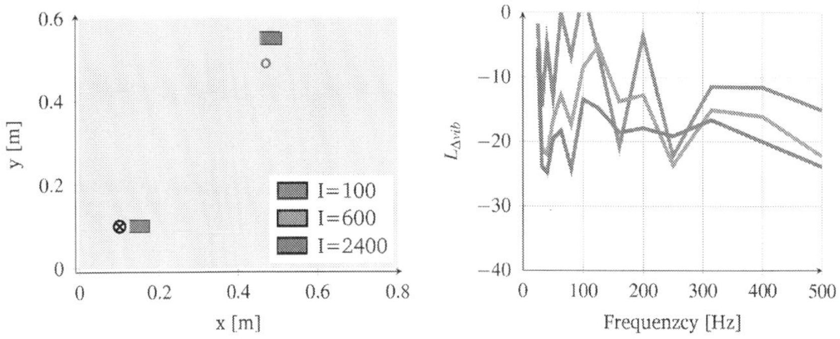

Figure 10. Optimised actuator positions for the causal system (left), vibration reduction at the error sensor in third octave bands (right).

In addition to the causality aspect, three different filter lengths (100, 600 and 2400) are used in the optimisation. The effects of a voltage limitation are investigated via the variation of the maximum actuator voltage: no limitation and 100 and 500 V limitations.

Figures 10 and **11** show the optimised actuators for the control systems with different filter lengths and different delays. By comparing **Figures 10** and **11** it can be easily seen that the actuator positions as well as the third octave vibration reduction differ significantly.

The optimised actuators of the NCS are located near the excitation for all filter weights. In contrast, the actuators of the CS are located near the error sensor except for the actuator optimised with a filter length of (I = 100). The CS actuators with a filter length of 600 and 2400 are nearly collocated, but the predicted performance differs by 5–10 dB in several third octave bands. This result matches with Section 3.5., where the influence of the filter weights on the vibration reduction is similar. Yet, it also shows the importance of including the filter length into an optimisation to improve the prediction accuracy. However, the different actuator placements need to be explained not only by the filter weights but rather by a combination of causality and filter weights.

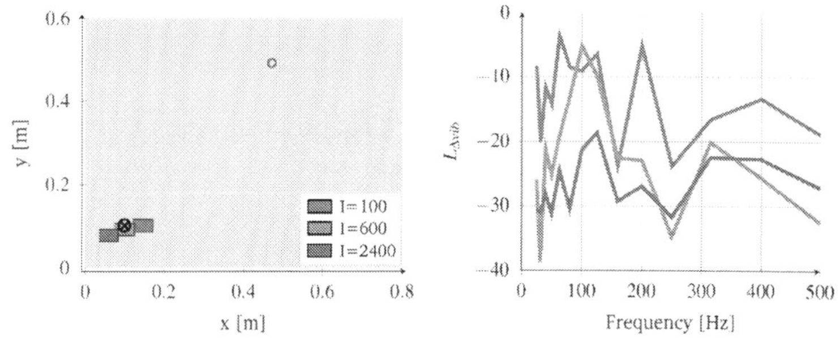

Figure 11. Optimised actuator positions for the non-causal system (left), vibration reduction at the error sensor in third octave bands (right).

To explain the actuator placement in more detail, the filter quality and the delays of the primary and secondary path are analysed. The filter qualities of the three actuator positions of the NCS are presented in **Figure 12** and those of the CS system in **Figure 13**. **Figure 11** shows different actuator positions of the NCSs despite the fact that all systems are causal due to the synthetic delay of 20 samples. So, the filter length has a major influence on the actuator placement which can be seen in **Figure 12**. The actuator position optimised with 100 filter weights has the highest filter quality of up to 400 filter weights, which means that the impulse response of the optimal filter has a short impulse response. So, it can be stated that a short filter leads to an actuator placement where a short impulse response of the optimal controller can be realised. This is the case if the secondary path is very similar to the primary path.

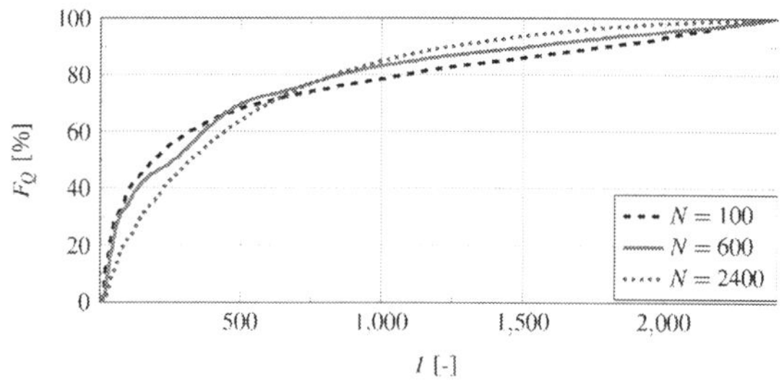

Figure 12. Filter qualities of the non-causal systems.

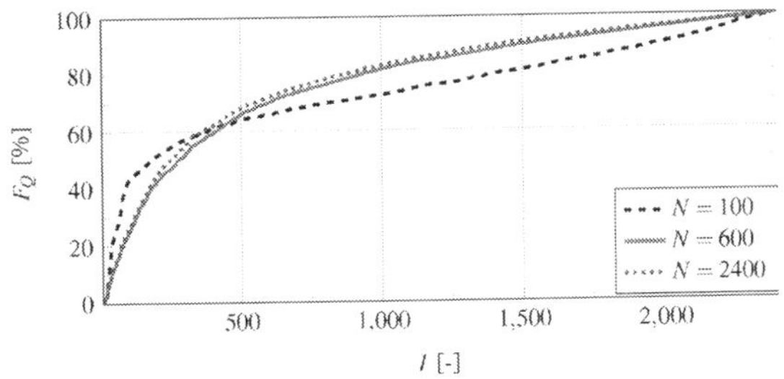

Figure 13. Filter qualities of the causal systems.

A comparison of the actuator positions of the CSs and the NCSs shows that the optimised actuator position for a short filter ($I = 100$) is the same for both systems. In contrast, the actuator positions for longer filters differ significantly between the two systems. To summarise, it can be said that the actuator position optimised with a short filter is not very sensitive to causality. Again, this matches Section 3.5., where it is shown that a time-advanced reference signal cannot be used for a system with a very short filter.

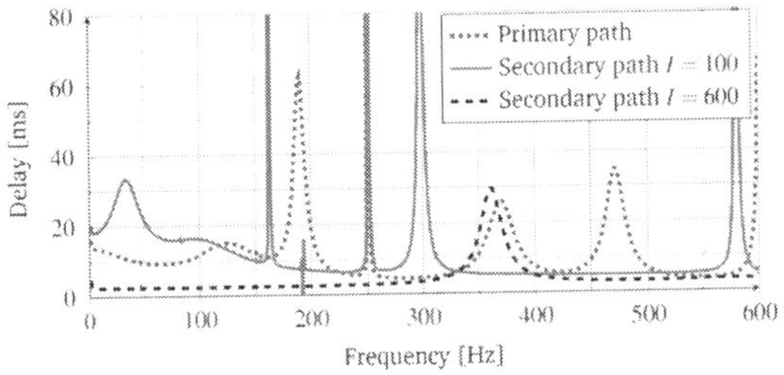

Figure 14. Group delays of the primary path and the causal system secondary paths.

For longer filters, the effect of the causality constraint becomes more pronounced. While the NCS was made causal by the synthetic delay induced in the primary path, the CS system is not causal if the actuator is located at the excitation point. So, in the case of the CS the causality constraint has to be fulfilled with a different actuator placement with smaller delays. This is achieved with the CS actuator placement with long filters. **Figure 14** presents the delays of the primary path and of the two secondary paths of the CSs (for $I = 100$ and 600).

By analysing the group delays it can be shown that the group delay of the actuator located near the error sensor has a smaller delay than the primary path. Only in the region of a non-minimum phase zero is the group delay higher than that of the primary path. In contrast, the actuator position near the excitation point has a larger group delay than the primary path nearly all over the frequency band, except some non-minimum phase zeros. The delay analysis clearly indicates that the causality constraint influences the actuator location for longer filters. Yet, because of the synthetic delay (17 ms) induced in the primary path, the causality constraint does not affect the NCS locations. If the primary delay is 17 ms larger than the delay presented in **Figure 14**, the causality constraint is fulfilled even for the actuator located at the excitation point.

The last parameter investigated in this section is the influence of structural damping. In the case of higher structural damping, the vibration energy is not localised in the resonant frequencies and the filter effect of the structure is smaller. Accordingly, the deterministic parts of the disturbance signal are reduced and a non-causal actuator location has an even worse performance. This can be seen in **Figure 15**, which presents the optimised actuator locations for the CSs.

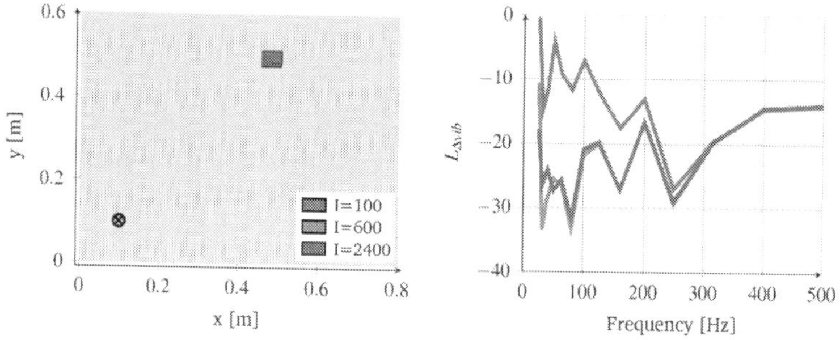

Figure 15. Optimised actuator positions for the causal system (left) with medium structural damping, vibration reduction at the error sensor in third octave bands (right).

In the case of medium structural damping, even the actuator optimised with 100 filter weights is located at the error sensor which is not the case for the light damping. This is due to the stronger performance penalties of the causality constraint for higher damped systems. A second tendency of the medium damped system can be seen on the right side of **Figure 15**. Compared to **Figure 10**, the performance for the higher structural damping is saturated with 600 filter weights. The need for less filter weights is due to the shorter impulse responses of the systems with higher structural damping.

4.3. Experimental validation

To experimentally validate the simulation results, two optimised actuator positions are realised on an aluminium plate. The experiments are focused on the different actuator positions caused by the variation of the filter length for the NCS configuration. The influence of the delays and the causality constraint was already shown by the author in [21] and is excluded in this investigation. For the experiments the actuator placements optimised with filter length of 100 and 2400 are realised.

The experiments are conducted on the test bed presented already in **Figure 1**. As it is presented in Section 3.2, a secondary path model and a synchronous measurement of the reference signal and the disturbance signal are required. Therefore, the piezoelectric patch actuator is driven by band-limited white noise and the signal of the accelerometer and the disturbance are measured simultaneously. The time series are post-processed with a sub-space-based identification algorithm which identifies a state-space model [30]. Afterwards, the excitation shaker is driven with a band-limited white noise signal. The reference signal is measured with a load cell between the plate and the shaker and the disturbance signal is measured simultaneously at the accelerometer. All sensor signals are low-pass filtered and fed into a rapid prototyping DSP. The excitation is also realised with the DSP and fed to reconstruction low-pass filters. In **Table 4**, a complete list of the experimental hardware is presented.

Hardware	Type	Additional information
DSP	dSpace® ds1006	Sampling frequency: 1200 Hz
Low-pass filters	KEMO Card Master 255G	Cut-off frequency: 560 Hz
Accelerometer	PCB 352A24	Mass: 0.8 g
Shaker	LMS 201	
Piezoelectric patch actuator	PI DuraAct P-876.A15 (50 × 30 × 0.5 mm active area)	Mass: 10 g

Table 4. Experimental hardware.

After the offline calculation of the optimal filter, the filter weights are implemented in the DSP and the controlled performance can be measured. **Figure 16** presents the performance of the two actuator positions where two filters with 100 and 2400 weights are designed.

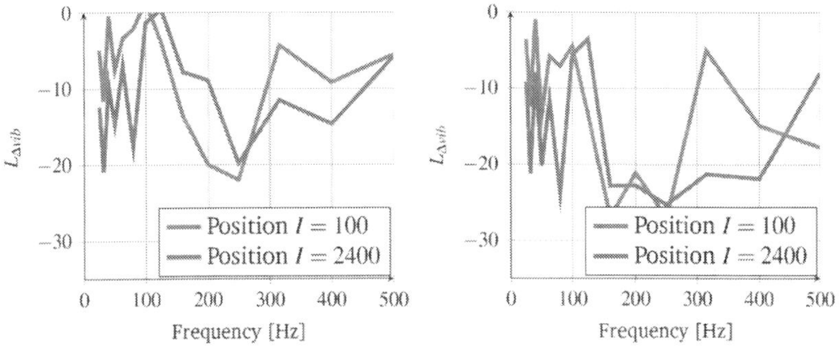

Figure 16. Experimentally measured vibration reduction in third octave bands of the two investigated actuator positions, with a filter length of 100 (left) and a filter length of 2400 (right).

In contrast to the simulation result, the actuator position optimised with 100 filter weights shows no performance advantages in comparison to the position optimised with 2400 filter weights. The opposite is the case for both filter lengths, the actuator position optimised with 2400 filter weights outperforms the position optimised with 100 filter weights. The total reduction of the feedforward controller using the actuator position optimised with 2400 filter weights integrated over the frequency band is 2 dB larger using in the case of an optimal filter with length 100 and 5 dB better using 2400 filter weights. This is indeed a surprising result.

This could be due to the model uncertainties shown in Section 2. Maybe, the optimisation with different filter weights is more sensitive to damping uncertainties. Another possible reason for this disagreement could be the simplified modelling of the piezoelectric patch actuators which maybe influence the performance of the feedforward controller especially in the lower frequency range. Nevertheless, the filter qualities for the different actuator positions fit the simulated ones. **Figure 17** shows that the actuator position optimised with 100 filter weights has a higher filter quality for short filters than the position optimised with 2400 filter weights.

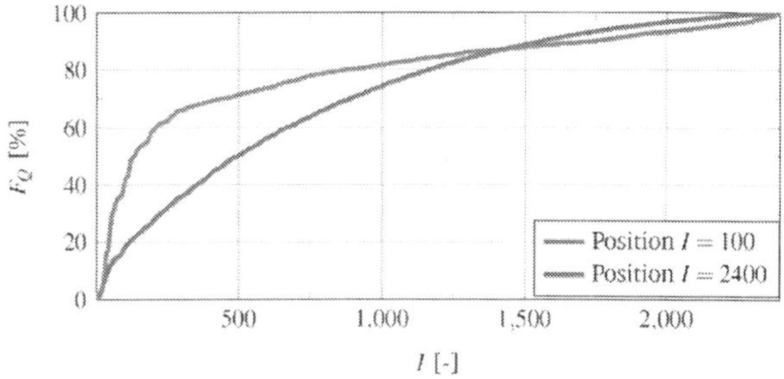

Figure 17. Filter qualities of the filters used in the experimental investigations.

In short, further investigations have to be done to investigate the dependency of the optimal actuator placement on the filter weights. In particular, the influences of the structural damping or other model uncertainties have to be shown.

5. Conclusion

This chapter presents and analyses the parameters influencing the performance of a feedforward control system. To this end, the basic theory of feedforward control systems is presented and the parameters (causality, FIR filter weights, saturation, etc.) which are necessary to design technical feedforward systems are studied. Furthermore, an extensive parametric study is presented in which the influence of each parameter is analysed. Therefore, each parameter is varied in a wide range. This chapter also shows the importance of modelling these parameters inside an actuator placement optimisation routine. During the optimisation, the influence of causality, the filter weights and the damping properties of the structure on the actuator placement is analysed.

It is shown that the causality constraint and the length of the FIR filter are of crucial importance for the feedforward control performance. A very short filter and large delays in the secondary path can significantly reduce the performance of a feedforward controller. Furthermore, the effect of structural damping on the performance is analysed. Under non-causal conditions, a larger structural damping leads to a decreased performance because the filtering effect of the structural resonances is reduced and so are, as a consequence, the deterministic properties of the disturbance signal. In addition, the effects of a limited actuator performance (i.e. saturation effects) and the use of an overdetermined (rectangular) system (i.e. more sensors than actuators) are studied. It is shown that saturation and over determinacy lead to a performance cut-off.

After the analysis of the parameters, an actuator placement optimisation is presented that incorporates the analysed parameters of a feedforward control system into the actuator placement. The cost function of the optimisation is chosen to be the flexural vibration at a fixed error sensor so as to ease the analysis of the optimisation result. It can be shown that the actuator position is dependent on the delays that determine the causality as well as on the filter length. For very short FIR filters, the actuator is located at positions at which the optimal control filter has a short impulse response and thus compensates the filter length. Only if the filter length is extended to a certain level, the causality aspect influences the actuator positions. The influence of the causality aspect on the actuator placement increases with a higher structural damping.

In the experiments, the simulated influence of the filter weights on the actuator positions cannot be validated. The filter qualities of the investigated actuator positions equals the ones that are simulated, but the overall performance are not in agreement with the simulations. Further studies should investigate if the model uncertainties are the reason for the deviations.

To summarise, it is shown that the design of a feedforward controller is highly dependent on the considered realisation. It is important to know which time-advance can be gathered by the reference signal (for causality aspects), which length of the FIR filter can be realised with the chosen digital signal processor, etc. An optimisation of the actuator and sensor positions is only effective if these aspects are known.

In future studies, the optimisation of actuator and sensor locations for global cost functions such as global vibration or sound power reduction will be investigated. When using global cost functions, further parameters need to be considered for the optimisation of sensor and actuator positions for feedforward control systems. Research should focus, for example, on the pinning effect and the number of sensors and actuators.

Author details

Thomas Haase

Address all correspondence to: thomas.haase@dlr.de

1 German Aerospace Center (DLR), Institute of Composite Structures and Adaptive Systems, Lilienthalplatz, Braunschweig, Germany

2 Technische Universität Braunschweig, Institut für Adaptronik und Funktionsintegartion (iAF), Langer Kamp, Braunschweig, Germany

References

[1] Padula S.L., Kincaid R.K. Optimization strategies for sensor and actuator placement. NASA Center for AeroSpace Information (CASI). 1999, 17 p.

[2] Frecker M.J. Recent advances in optimization of smart structures and actuators. Journal of Intelligent Material Systems and Structures. 2003;14:207–216.

[3] De Fonseca P., Sas P., Van Brussel H. A comparative study of methods for optimising sensor and actuator locations in active control applications. Journal of Sound and Vibration. 1999;221:651–679.

[4] Gupta V., Sharma M., Thakur N. Optimization criteria for optimal placement of piezoelectric sensors and actuators on a smart structure: a technical review. Journal of Intelligent Material Systems and Structures. 2010;21:1227–1243.

[5] Clark R.L., Fuller C.R. Optimal placement of piezoelectric actuators and polyvinylidene fluoride error sensors in active structural acoustic control approaches. The Journal of the Acoustical Society of America. 1992;3:1521–1533.

[6] Gawronski W.K. Advanced Structural Dynamics and Active Control of Structures. 1st ed. New York: Springer; 2004.

[7] Wang B.-T. Optimal placement of microphones and piezoelectric transducer actuators for far-field sound radiation control. The Journal of the Acoustical Society of America. 1995;5:2975–2984.

[8] Nijhuis M.O., de Boer A. Optimization strategy for actuator and sensor placement in active structural acoustic control. In: International Symposium on Active Control of Sound and Vibration, ACTIVE, July 15–17, 2002, Southampton, UK (pp. 621–632).

[9] Kim J., Varadan V.V., Varadan V.K. Finite element-optimization methods for active control of radiated sound from a plate structure. Journal of Smart Materials and Structures. 1995;4:318–326.

[10] Moshrefi-Torbati M., Keane A., Elliott S.J., Brennan M., Anthony D., Rogers E. Active vibration control (AVC) of a satellite boom structure using optimally positioned stacked piezoelectric actuators. Journal of Sound and Vibration. 2006;292:203–220.

[11] Montazeri A., Poshtan J. Optimizing a multi-channel ANC system for broadband noise cancellation in a telephone kiosk using genetic algorithms. Shock and Vibration. 2009;16:241–260.

[12] Belgacem W., Berry A., Masson P. Active vibration control on a quarter-car for cancellation of road noise disturbance. Journal of Sound and Vibration. 2012;331(14):3240–3254.

[13] Kuo S.M., Morgan D.R. Active Noise Control System: Algorithms and DSP Implementations. 1st ed. New York: John Wiley & Sons, Inc.; 1996.

[14] Nelson P.A., Elliott S.J. Active Control of Sound. London: Academic Press; 1991.

[15] Burdisso R., Vipperman J., Fuller C. Causality analysis of feedforward-controlled systems with broadband inputs. The Journal of the Acoustical Society of America. 1993;94:234–242.

[16] Haase T., Algermissen S., Unruh O., Misol M. Experiments on active control of counter-rotating open rotor interior noise. Acta Acoustica United with Acustica. 2014;100:448–457. DOI: 10.3813/AAA.918725.

[17] Miller S., Shipps J.C. Active noise cancelling muffler. US-Patent. 1998; Patent Number: US5748749A.

[18] Hammerschmidt C. Active noise cancellation applied to diesel exhaust, 2010, [Internet]. Avaiable form: : http://www.eetimes.com/document.asp?doc_id=1257672 [Acessed: 22.06.2016].

[19] Elliott S.J. Signal Processing for Active Control. 1st ed. London: Academic Press; 2001. 511 p.

[20] Widrow B., Stearns S., Stearns P.N. Adaptive Signal Processing. 1st ed. Englewood Cliffs, NJ: Prentice Hall; 1985. 485 p.

[21] Haase T., Misol M., Rose M. Optimal placement of flat piezoceramic actuators for feedforward systems under the influence of real-time hardware delays. Journal of Sound and Vibration. 2015;345:34–46.

[22] Janocha H., Liu B. Simulation approach and causality evaluation for an active noise control system. IEE Proceedings in Control Theory and Applications; IET. 1998;145(4): 423–426.

[23] Minkoff J. The operation of multichannel feedforward adaptive systems. IEEE Transactions on Signal Processing. 1997;45(12):2993–3005.

[24] Rose M. Modal based correction methods for the placement of piezoceramic Modules. In: ASME International Mechanical Engineering Congress and Exposition. 05 November 2005; Orlando, Florida, USA; 2005.

[25] Lee K.-A., Gan W.-S, Kuo S.M. Subband Adaptive Filtering: Theory and Implementation. 1st ed. West Sussex, UK: John Wiley & Sons Ltd.; 2009.

[26] Pohlheim H. Evolutionäre Algorithmen. 1st ed. New York: Springer; 2000. DOI: 978-3-642-57137-4.

[27] Pohlheim H. GEATbx: Genetic and evolutionary algorithm toolbox for use with MATLAB [Internet]. 2006. Available from: http://www.geatbx.com/docu/ [Accessed: 10.11.2011].

[28] Unruh O., Haase T., Pohl M. Application of a load-bearing passive and active vibration isolation system in hydraulic drives. In: MOVIC/RASD Conference Proceedings; 3–6 July 2016; Southampton, UK; 2016.

[29] Kessissoglou N.J., Ragnarsson P., Löfgren A. An analytical and experimental comparison of optimal actuator and error sensor location for vibration attenuation. Journal of Sound and Vibration. 2002;260:671–691.

[30] Kateyama T. Sub-space based identification algorithms. 1st ed. New York: Springer; 2005.

[31] Wang W.Y., Yue K., Shi D.Y. Optimal research of actuator placement for piezoelectric smart structure. Journal of Key Engineering Materials. 2010;419:173–176.

[32] Preumont A. Vibration Control of Active Structures. 1st ed. Dordrecht: Kluwer Academic Publishers; 1997.

[33] PI Ceramic. DuraAct Flaechenwandler [Internet]. 2015. Available from: http://www.piceramic.de/produkt-detailseite/p-876-101790.html [Accessed: 02.02.2015].

Chapter 4

AC Variable-Speed Drives and Noise of Magnetic Origin: A Strategy to Control the PWM Switching Effects

Jean-François Brudny, Fabrice Morganti and Jean-Philippe Lecointe

Additional information is available at the end of the chapter

http://dx.doi.org/10.5772/64520

Abstract

The presented developments concern the definition of a method to characterize simply the pulse width-modulated (PWM) switching three-phase harmonic systems. This characterization, which makes it possible to control the components generated by the switching anticipating the negative effects possibly generated, is exploited in the context of reducing the noise of magnetic origin that can affect the operating of AC variable speed drives.

Keywords: AC variable speed drives, noise of magnetic origin, radial vibrations, PWM inverters, switching control

1. Introduction

The use of Pulse Width-Modulated (PWM) inverters is widespread since many years in the field of variable speed AC motor drives. The main advantage, in addition to the offered flexibility and the implementation easiness, concerns the rejection at high frequencies of the voltage switching three-phase harmonic systems (STPHSs). This naturally leads to minimizing their effects considering the inductive machine behavior at these frequencies. The beneficial aspects are real for the torque harmonics [1, 2] insofar as the concerned mechanical resonances, which are characterized by frequencies varying from a few tens to a few hundred Hz, can amplify the effects of low frequency components as, for example, six-step inverters. However, other phenomena, particularly important, may occur. The latter concern the iron [3, 4] and copper losses, the noise of magnetic origin [5, 6] but also premature aging of the electrical winding insulation [7, 8]. To overcome these effects, a simple solution consists of introducing a filter connected to the inverter outputs. However, taking into account the additional costs required by this solution that also increases the size of the "inverter," this technique is not always compatible with the industrial constraints. That is why research teams have proposed other solutions to control some of these effects. These last can be grouped into three categories. The first acts on the power electronic converter architecture in order to minimize the voltage variation amplitude during a switching leading, for example, to the multilevel converters [9]. The second concerns the machine design [10], by introducing, for example, specific dampers [11, 12] to reduce the switching air gap flux density harmonics. The third concerns the definition of specific control strategies.

The study presented in this chapter is in the third category, with the goals of reducing the noise of magnetic origin. Many references dealing with this particular subject appear in the scientific literature [13–18]. Realized developments present the originality to remove a force component whose effects are particularly troublesome if that component excites a stator mechanical resonance that may be located at frequencies of a few kHz which correspond, in terms of human hearing response, to a particularly sensitive area [19]. Let us assume that this force component is generated by a given balanced STPHS in the output signals of the converter. The principle consists in removing this STPHS by making homopolar an intermediate STPHS which is at its origin. This principle, which requires to use three distinct carriers, has been already presented [20–22]. It is based primarily on the STPHS phase angle determination. To facilitate this determination, a Sinus—Sinus (SS) PWM is associated to the classic Sinus—Triangle (SΔ) PWM. It has also been shown that this suppression has a drawback insofar it is accompanied by the appearance of certain harmonics, which are in some cases unbalanced, that may challenge the proposed principle. To minimize the negative effects of this procedure, a strategy called "carrier-phase jump" [22] is presented, it consists of removing alternately two STPHSs from the output signals, exploiting the additional degree of freedom provided by this technique. Then, a procedure that uses carriers defined with different frequencies is disclosed.

This chapter is therefore mainly devoted to the presentation and the experimental implementation of the methods just mentioned, considering a symmetrical, "p" pole pair induction machine (IM) whose magnetic circuit is supposed to be nonsaturated. The used inverter has a basic structure: two levels and triangular carriers. As the inverter generates voltage STPHSs, some of them being unbalanced, the first part concerns the characterization of the acoustic noise generated by the IM in these conditions. Such an analysis seems a priori original as the state-of-the-art shows that no study have been proposed with such an approach. The second part shows the method to characterize simply the STPHSs, especially their harmonic-phase sequences. The third part presents the application of the method and the control strategy applied to the inverter in order to act on STPHSs. The possible negative effects of the control are also described. In order to minimize these effects, the fourth part exposes the carrier-phase jump control that has not been published in a journal. The fifth part presents, for the first time, the method that consists in using different frequencies for the three carriers. The last part is devoted to the presentation of experimental results concerning IM magnetic noise and radial vibration reductions.

A time "k" rank harmonic of a variable "z" on phase "q" (q = 1, 2 or 3) will be denoted z_q^k. Let us point out that upper indexes "(s)" and "(Δ)" will also be used to distinguish the variables according to the sinusoidal or triangular carrier waveforms. No index means that the relations are valuable whatever the case. The STPHSs will be characterized not only by their frequency but also with the order of the phase sequence: Clockwise (C), Anticlockwise (A), and Homopolar (H).

2. Magnetic noise components

The presentation of the proposed strategy requires to fully understand the origins of the various force components that produce the mechanical deformations of the stator massive external housing. This justifies the developments that are presented in this section. They are based on the definition of the "b" air gap flux density harmonic content. As saturation is neglected, "b" is substantially independent of the IM load conditions and, consequently, is defined by the product [23, 24]:

$$b = \varepsilon \Lambda \tag{1}$$

ε is the air gap magnetic potential difference generated by the stator at no load and Λ models the air gap per unit area permanence. The power supply fundamental frequency is denoted by "f" (angular frequency ω, period T). Regarding Λ, the relationship established in Ref. [25] is used:

$$\Lambda = \sum_{k_s}\sum_{k_r} \Lambda_{k_s k_r} = \sum_{k_s}\sum_{k_r} \hat{\Lambda}_{k_s k_r} \cos[(k_s N^s + k_r N^r)\alpha - k_r N^r \mathcal{D}] \tag{2}$$

N^S and N^r define the stator and rotor salience numbers (teeth). k_s and k_r are integers varying from $-\infty$ to $+\infty$. Let us consider a d^s stator spatial reference that is assumed to be confounded with the Phase 1 axis. α makes it possible to locate any point of the air gap with respect to d^s and D characterizes the rotor position relative to the same reference. Assuming that the rotor speed is imposed by (C) fundamental three-phase voltage system, it can be written: $\mathcal{D} = (1-s)\theta/p + \mathcal{D}_0$, with "s" the IM slip, $\theta = \omega t$ and $\mathcal{D}_0$ the $\mathcal{D}$ value for t = 0. A wave with a positive direction of rotation will be considered in the trigonometric direction, like the fundamental wave. A negative direction characterizes a wave rotating in the opposite direction. The permanence expression given by the relationship (2) is composed of:

- a constant term Λ_{00} which characterizes the air gap corrected thickness used to design the IMs,

- terms which depend on the stator toothing $\Lambda_{k_s 0}$,

- terms which depend on the rotor toothing Λ_{0k_r},

- terms $\Lambda_{k_s k_r}$ (for k_s and $k_r \neq 0$), which model the interaction between stator and rotor toothings.

In order to fix the approximated values of the terms, depending on the toothing, it can be noted that $\Lambda_{k_s 0}$, Λ_{0k_r} and $\Lambda_{k_s k_r}$ are, respectively, inversely proportional to $|k_s|$, $|k_r|$ and $|k_s k_r|$.

The expression (2), in its formulation, is similar to expressions given by other authors. Alger [23] considers the same relation but with only the fundamental of each group of terms ($k_s = k_r = 1$). Timar [24] considers the harmonic components with $\Lambda_{k_s 0}$ and Λ_{0k_r} but neglects the terms $\Lambda_{k_s k_r}$ by arguing that their contribution is low, which is erroneous when the slotting resonance phenomenon appears [25].

2.1 Case of balanced STPHSs

Assuming that (H) systems do not exist in the v_q voltages applied to the stator windings, "k" can be defined as a positive or negative integer that characterizes (C) and (A), respectively, balanced systems so that:

$$\varepsilon = \sum_k \sum_h \varepsilon_h^k = \sum_k \sum_h \hat{\varepsilon}_h^k \cos[k\theta - hp\alpha - \varphi_h^k] \tag{3}$$

"h" defined by: h = 1 + 6n with "n" an integer varying from $-\infty$ to $+\infty$, defines the rank of the space harmonics related to the nonuniform distribution of the winding wires at the stator frame. α is an angular abscissa for positioning any point of the air gap with respect to d^s. Let us point out that $\hat{\varepsilon}_h^k$ is inversely proportional to "h" and also to "k" because the RMS values of the i_q^k currents which define $\hat{\varepsilon}_h^k$ decrease generally with "k".

It follows that "b" is expressed by:

$$b = \sum_k \sum_h \sum_{k_s} \sum_{k_r} b_{hk_s k_r}^k \tag{4}$$

This "b" relationship leads to the "F" Maxwell force per unit area acting on the stator inner frame. Noting μ_0 the air permeability ($\mu_0 \approx 4\pi 10^{-7} H/m$), "F" is defined by:

$$F = \frac{1}{2\mu_0}\left[\sum_k \sum_h \sum_{k_s} \sum_{k_r} b^k_{hk_s k_r}\right]^2 \tag{5}$$

leading to express, in general [22], "F" by:

$$F = \sum_K \sum_H \sum_{K_s} \sum_{K_r} \hat{F}^K_{HK_s K_r} \cos[\Pi^k_{K_r}\theta - P_{HK_s K_r}\alpha + \varphi^K_{HK_r}] \tag{6}$$

These force components are very numerous, but all of them do not generate magnetic noise. In fact, the components to be considered must be characterized by:

- a $\hat{F}^K_{HK_s K_r}$ sufficient amplitude,

- a relatively small $P_{HK_s K_r}$ mode number ($P_{HK_s K_r} < 8$),

- a frequency $\Pi^k_{K_r} f$ close to a stator frame natural frequency.

As the sums imply several variables, it is obvious that the analytical expressions of the force components become heavy. As a result, to make the study easier, the phase angles will not be considered so that the characteristics will concern the frequencies and the pole numbers of the force components. Moreover, the components will be selected by taking into account only the forces which may have sufficient amplitudes. This approach allows taking into account the fact that the model given by Eq. (2) for characterizing the toothing, relies on hypotheses that decrease the reliability of the determination concerning the flux density components, and thus the force components, of very low amplitudes.

This approach is so justified in view of the criteria which characterize an annoying force. Indeed, it is shown that a force of high amplitude may have any effect on the noise although a force component with a much lower amplitude can be particularly important in terms of noise if its frequency is close to a natural resonance of the stator stack. However, at this level of the study, another difficulty concerns the transfer function for calculating the radial vibration amplitudes due to the force components. Analytical expressions have been proposed [23, 26], but the models are relatively inaccurate; the error on the vibration amplitudes, and so on the noise, may be important. The explained considerations reinforce our approach to characterize "F"; it consists in not systematically calculating the formal value of the amplitudes. In these conditions, introducing the following quantities:

$$\left.\begin{aligned} &S = N^r(1-s)/p \\ &hp_{(+)} = hp + (k_s N^s + k_r N^r) \\ &hp_{(-)} = hp - (k_s N^s + k_r N^r) \\ &\hat{b}^k_{hk_s k_r} = \hat{\varepsilon}^k_h \hat{\Lambda}_{k_s k_r} / 2 \end{aligned}\right\} \tag{7}$$

It can be written:

$$b^k_{hk_s k_r} = \hat{b}^k_{hk_s k_r} \left\{ \cos[(k - k_r S)\omega t - hp_{(-)}\alpha] + \cos[(k + k_r S)\omega t - hp_{(+)}\alpha] \right\} \tag{8}$$

The component selection is done from the development of the square of expression (8), considering the previous criteria about the force amplitudes. Considering expression (5), it appears that some force terms result from the square of each flux density term, the others force components depending on the double products between the flux density components. The numerical calculations [25] show, considering the fundamental component, that $\hat{b}^1_{100} >> \hat{b}^1_{hk_s k_r}$ for every combination "$hk_s k_r$" different from "100". For example, for a classical machine, for $\hat{b}^1_{100} = 1T$, $\hat{b}^1_{hk_s k_r}$ is about 0.02 T for the highest values which correspond to the first values of the parameters that intervene in the inferior combination. It means that concerning the squared terms, only $(b^1_{100})^2$ has to be considered because, for a combination inferior different from "100," the other squared terms leads to terms of which amplitudes, in the most favorable case, are of the order of 10^4 lower than $(\hat{b}^1_{100})^2$. Concerning the double products, the previous example shows that only the terms which include $\hat{b}^1_{100}$ have to be considered.

By extending this particularity to the terms which concern harmonics of "k" rank, it appears that "F" given by expression (5) can also be written as:

$$F^* = \frac{1}{2\mu_0} \sum_k \sum_h \sum_{k_s} \sum_{k_r} b^k_{100} b^{k'}_{h'k'_s k'_r} \tag{9}$$

The use of F* instead of F distinguishes the force defined by the rules of art or according to our agreement to simplify the mathematical developments with an impact mainly on the "F" component amplitudes.

Introducing the parameters k', h', k'_r and k'_s, that have the same functionality as k, h, k_s and k_r, makes it possible to distinguish two terms, generally different, to exploit the sums defined by expressions (4) and (5). On the other hand, if for a "k" given value, k', h', k'_r and $k'_{s'}$ take all the values in their variation domain, the expression (9) takes also into account the force components which result from the squared terms. Thus, to characterize "F^*," a relationship similar to expression (6) will be used, by adapting the coefficients of expression (6). It can be written:

$$F^* = \sum_k \sum_h \sum_{k_s} \sum_{k_r} F^{k,k'}_{h'k'_s k'_r} = \sum_k \sum_h \sum_{k_s} \sum_{k_r} \hat{F}^{k,k'}_{h'k'_s k'_r} \cos[\Pi^{k,k'}_{h'k'_s k'_r}\theta - P^{k,k'}_{h'k'_s k'_r}\alpha] \quad (10)$$

The use of (9) allows to do a systematic sum without any analysis but with the drawback of a high uncertainty concerning the amplitudes. However, this uncertainty is not of importance insofar a selection of the component was made and, on the other hand, this uncertainty may be of the same order as the fact to neglect the phase angles. In return, the analytical expressions are notably simplified, but with a strict respect of the frequencies and the pole numbers. Developing the product $b^k_{100} b^{k'}_{h'k'_s k'_r}$ leads to show that $F^{k,k'}_{h'k'_s k'_r}$ is composed of four terms whose coefficients $\Pi^{k,k'}_{h'k'_s k'_r}$ and $P^{k,k'}_{h'k'_s k'_r}$ are given in **Table 1**. Their exploitation requires to consider first the upper signs and then the lower ones.

	$\Pi^{k,k'}_{h'k'_s k'_r}$	$P^{k,k'}_{h'k'_s k'_r}$	$\Pi^{k,k'}_{h'k'_s k'_r}$	$P^{k,k'}_{h'k'_s k'_r}$
$b^k_{100} b^{k'}_{h'k'_s k'_r}$	$k + k' \mp k'_r S$	$p + h'p_{(\mp)}$	$k - k' \pm k'_r S$	$p - h'p_{(\mp)}$

Table 1. Characterization of the $b^k_{100} b^{k'}_{h'k'_s k'_r}$ coefficients.

At this stage of our developments, two cases may be considered according to the "k" values. Let us consider the relationship: $v^k_q = \frac{d\psi^k_q}{dt}$, where ψ^k_q corresponds to phase "q" linked flux resulting from air gap magnetic effects generated by the "k" rank three-phase harmonic voltage. It can be shown, for given "f," that $\hat{b}^k_{100}$ is proportional to $\hat{v}^k_q$ and inversely proportional to "k". Let us assume that $\hat{v}^k_q$ is same order of magnitude as $\hat{v}^1_q$.

- For low "k" values (first odd values), $\hat{b}^k_{100}$, for k≠1, is practically the same order of magnitude as $\hat{b}^1_{100}$. This means that the relation (10) should be used as presented.

- For high "k" values such as those that characterize STPHSs, $\hat{b}^{k}_{100}$, for k≠1, is very small comparing $\hat{b}^{1}_{100}$. For example, considering the case of a supply via a PWM inverter, for a switching frequency of 50 times greater than "f," the largest amplitude of these harmonic components, considering the previous equality on the voltages, is in the range of 0.02 T for $\hat{b}^{1}_{100} = 1T$. Given the comments that were made earlier about the amplitudes of the components resulting from the slotting effect, it can be deduced that the squared amplitudes of flux density components generated by the switching, for the most important, are of the order of 10^4 lower than $(\hat{b}^{1}_{100})^2$. Therefore, the effects of teeth associated with these quantities may be neglected. Then, Eq. (9) can be written as follows:

$$F^* = \frac{1}{2\mu_0}\left\{\sum_h\sum_{k_s}\sum_{k_r} b^{1}_{100} b^{1'}_{hk_sk_r} + \sum_k b^{k}_{100} b^{k'}_{100}\right\} \tag{11}$$

leading to define F* by the relationship:

$$F^* = \sum_h\sum_{k_s}\sum_{k_r} \hat{F}^{1,1}_{hk_sk_r}\cos[\Pi^{1,1'}_{hk_sk_r}\theta - P^{1,1'}_{hk_sk_r}\alpha] + \sum_k\sum_{k_r} \hat{F}^{k,k'}_{100}\cos[\Pi^{k,k'}_{100}\theta - P^{k,k'}_{100}\alpha] \tag{12}$$

whose coefficients are given in **Table 2**.

	$\Pi^{k,k'}_{hk_sk_r}$	$P^{k,k'}_{hk_sk_r}$	$\Pi^{k,k'}_{hk_sk_r}$	$P^{k,k'}_{hk_sk_r}$
$\hat{b}^{1}_{100}\hat{b}^{1}_{hk_sk_r}$	$2 \mp k_rS$	$p + hp_{(\mp)}$	$\pm k_rS$	$p - hp_{(\mp)}$
$\hat{b}^{k}_{100}\hat{b}^{k'}_{100}$	k+k'	2p	k-k'	0

Table 2. Characterization of the coefficients which define F* for high "k" values.

Considering Eq. (10), the square of the rank "k" term is obtained for $k = k'$ by considering the inferior combination "100" for "h', k'_s, k'_r". This also corresponds, considering Eq. (12), to exploit the coefficients of the second row of **Table 2** with k' = k. It results that these squares lead to two force components which can be identified with $\hat{F}^{k,k}_{100}\cos[2k\theta - 2p\alpha]$ and $\hat{F}^{k,k}_{100}$. The first term is a rotating wave with a speed $k\omega/p$ and with a direction positive or negative according to the sign of "k". The second term leads to a permanent deformation of the IM external housing with a constant amplitude. This force component does not generate any vibration and so any noise. It is qualified as stationary force.

2.2. Case of unbalanced STPHSs

It is supposed that, even for this case, there is no (H) systems that is justified in section III. In order to distinguish, for a "k" given value, the kf frequency (C) and (A) systems that appear with the imbalance, the ranks k_C and k_A are introduced. Previously, this distinction was not necessary insofar as harmonic phase system was either (C) or (A), implicitly, "k" so take either the value k_C or k_A. In this case, for certain values of "k," ε_h^k is expressed as follows:

$$\varepsilon_h^k = \hat{\varepsilon}_h^{k_C}\cos[k_C\theta - hp\alpha] + \hat{\varepsilon}_h^{k_A}\cos[k_A\theta - hp\alpha] \tag{13}$$

The single goal of introducing k_C and k_A is to distinguish the systems knowing that the absolute values of k, k_C and k_A are equal, but the amplitudes of the components relative to k_C and k_A are generally different. If, before the imbalance, "k" was positive, then k_C is positive and k_A is negative. If, on the contrary, before the imbalance "k" was negative, then $k_C < 0$ and $k_A > 0$. As the definition of "h" is not changed, it appears that, for $h > 0$ and $k < 0$, the mmf component depending on k_A corresponds in fact to a (C) system.

The relationship (9) does not change, only the set of possible values of "k" is enriched following the splitting of certain harmonic three-phase systems. It follows that F* is still given by Eqs. (10) or (12) so that **Tables 1** and **2** remain valid. Let us point out finally that the expression (8) that gives $b_{hk_sk_r}^k$, assuming that the rank "k" three-phase system is affected by an imbalance, as: $k = k_C = -k_A$, becomes:

$$\begin{aligned} b_{hk_sk_r}^k &= b_{hk_sk_r}^{k_C} + b_{hk_sk_r}^{k_A} = \hat{b}_{hk_sk_r}^{k_C}\left\{\cos[(k-k_rS)\omega t - hp_{(-)}\alpha] + \cos[(k+k_rS)\omega t - hp_{(+)}\alpha\right\} + \\ &+\hat{b}_{hk_sk_r}^{k_A}\left\{\cos[(k+k_rS)\omega t + hp_{(-)}\alpha] + \cos[(k-k_rS)\omega t + hp_{(+)}\alpha]\right\} \end{aligned} \tag{14}$$

The force components which result from $(\hat{b}_{100}^k)^2$, $\hat{b}_{100}^k$ being deduced from Eq. (14), lead to a denser harmonic content compared to the balanced STPHS case. It is composed of:

- a term of the form "$\cos(2k\omega t - 2p\alpha)$" whose amplitude is proportional to $(\hat{b}_{100}^{k_C})^2$
- a term of the form "$\cos(2k\omega t + 2p\alpha)$" whose amplitude is proportional to $(\hat{b}_{100}^{k_A})^2$
- a term of the form "$\cos 2k\omega t$" whose amplitude is proportional to the products $\hat{b}_{100}^{k_C}\hat{b}_{100}^{k_A}$

- two stationary terms whose amplitudes are proportional to $(\hat{b}^{k_C}_{100})^2$ and to $(\hat{b}^{k_A}_{100})^2$,

- two "pseudo-stationary" terms of the form "$\cos 2p\alpha$" whose amplitudes are proportional to $\hat{b}^{k_C}_{100}\hat{b}^{k_A}_{100}$.

The pseudo-stationary state differs from the stationary state by the fact that the deformation of the stator housing, independent of the time, is nonuniform. Such components do not generate acoustic noise as the stationary components.

The $\cos 2k\omega t$ terms characterize force components with any pole. Thus the deformation of the external housing is uniform with an amplitude varying with time. This kind of deformation is qualifying of "breathing mode". Numerical applications will show the acoustic impact of such a force.

Concerning these developments, simplified expressions to characterize the force components in terms of frequencies and polarities, by eliminating components that exhibited too low amplitudes to cause significant effects, have been suggested. However, in terms of simplification, considerations about the pole numbers of the forces can significantly reduce the number of terms to be considered to characterize "F*". For example, it is immediately apparent, when k_s and k_r are nonzero and of the same sign, the absolute values of the sums which characterize the pole numbers given by (7) are very high and do not comply, therefore, the conditions for retaining the corresponding force components. This will be developed in the next section in relation to numerical applications that concern a three-phase IM with p = 2, $N^S = 36$, and $N^r = 24$.

2.3. Numerical applications

2.3.1. *Practical evaluation of force components*

- Assuming that all the STPHSs are balanced, for "h" successively equals to 1, −5, 7, −11, 13, −17, 19, −23, and 25, and k_s and k_r varying between −9 and +9, the number of force components which have mode numbers lower than 8 has been determined from Eq. (7). This analysis shows that all these components are characterized by $hp_{(+)} = hp_{(-)} = p = 2$.Whatever "$h$," for a given value of k_s, two force components appear. They are characterized by different values of k_r, except for $h = 1$ for which these two values of k_r are the same leading, because of the particularity concerning the pole numbers, to identical expressions. The number of components to consider according to the values taken by "h" is presented in **Table 3**.

- The second criterion concerns the amplitudes by considering $|hk_sk_r|$ and the terms for which these quantities are lower than 100. Practically, as for the fundamental this quantity is identified to 1, by supposing in first approximation that the force amplitudes are inversely

h	1	−5	7	−11	13	−17	19	−23	25
Criteria: force pole numbers	14	12	12	12	12	14	14	12	12
+ Criteria on the amplitudes	14	8	8	6	6	4	4	4	4

Table 3. Numbers of force components for given k and k′.

proportional to $\left|hk_sk_r\right|$ (it is a very optimistic consideration for the amplitudes), it means that the components whose amplitudes are lower than one hundredth of the fundamental amplitude are neglected. When $|h|$ increases, the number of components, which satisfy the two conditions, decreases. The second row in **Table 3** gives the state of remaining components.

- For given values of k and k', because of the particularity of the force pole numbers, as the frequencies depend only on k_r as shown in **Table 1**, it is possible to group all the terms with the same k_r. That way, only 15 distinct components appear as shown in **Figure 1** with gives the theoretical spectrum. The latter shows, the amplitude variations with k_r. These amplitudes are defined by $1/\left|hk_sk_r\right|$ by taking into account, for each value of k_r, only the combination defined for the lowest value of $\left|hk_sk_r\right|$. It can be noted that no component exist for $k_r = \pm 8$ and ±7. At last, it appears that, for the considered case, the number of components is relatively small while covering rather a wide spectrum. For $k = k' = 1$ (machine supplied with sinusoidal waves), the frequencies deduced from **Table 1** are given by $(1 \mp k_rS)f$. At no load, for f = 50 Hz, the "s" slip is 0.2%, then S = 11.976. As a result, the spectrum is spread over a frequency band between 0 and 5439 Hz. Assuming, as above, a switching frequency equal to 50 f, the noise components caused by the teeth for the voltage component at 50 Hz will interfere with the noise generated by switching around 2500 Hz and, to a lesser extent, around 5000 Hz. For higher frequencies, only the effects generated by STPHSs will be visible.

2.3.2. *Impact of an imbalance on the noise measurements*

The main practical problem results from nonstationary force components which come from the squares of the terms following the imbalance. Let us denote "B" and "U" quantities contained in Eq. (12) which correspond, respectively, for balanced and unbalanced system. They are representative, to a constant, of the force components which act on the inner periphery of the stator.

$$\left.\begin{aligned} B &= (\hat{b}_{100}^{k})^2 \cos(2k\omega t - 2p\alpha) \\ U &= (\hat{b}_{100}^{k_C})^2 \cos(2k\omega t - 2p\alpha) + (\hat{b}_{100}^{k_A})^2 \cos(2k\omega t + 2p\alpha) + \hat{b}_{100}^{k_C}\hat{b}_{100}^{k_A} \cos 2k\omega t \end{aligned}\right\} \quad (15)$$

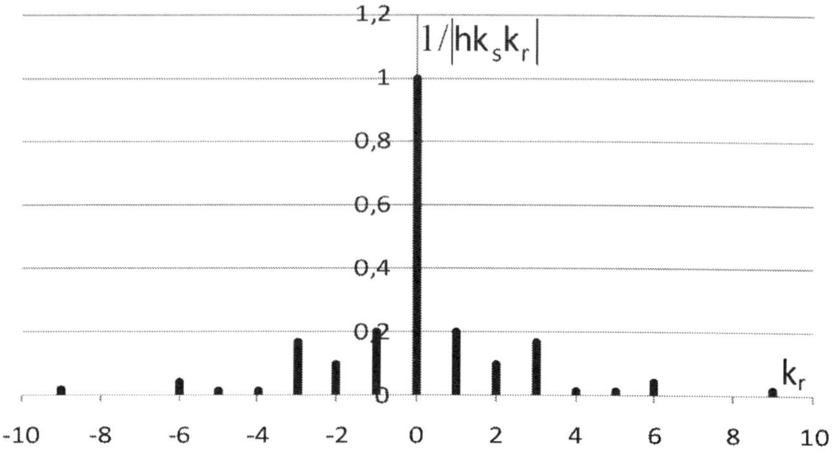

Figure 1. Theoretical frequency components generated by the slotting effect.

As "k" only affects the wave speeds, **Figure 2** presents for k = 1 and α taking successively the values 0, $\pi/4$ et $\pi/3$rd, the variations with time of B (**Figure 2b**) for $\hat{b}^{k}_{100} = 1T$ and of U (**Figure 2c**) for $\hat{b}^{k_C}_{100} = 0.55T$ and $\hat{b}^{k_A}_{100} = 0.45T$.

One can note, whatever the position of a microphone or an accelerometer around the IM, the same effects are measured when the STPHS is balanced. When the STPHS is unbalanced, for some positions ($\alpha = \pi/3$rd for example), there is practically no effect while at other places ($\alpha =$ 0rd, for example) these effects are particularly significant.

That is why it is necessary to do measurements with at least two devices spatially shifted to adequately characterize the effects, the shifting being dependent on the wave deformation mode numbers.

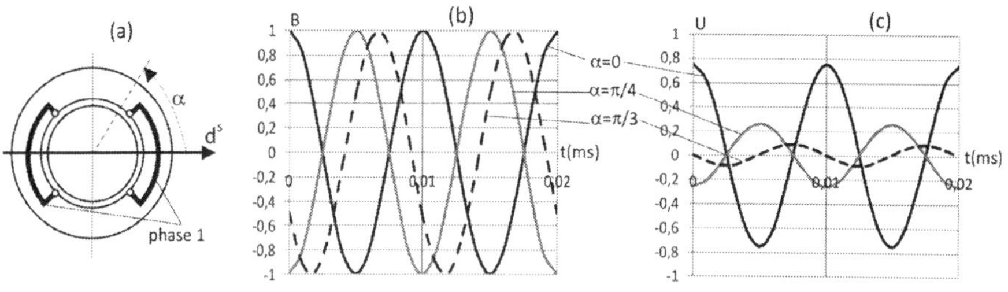

Figure 2. Forces acting on the inner stator surface: (b) balanced STPHS, (c) unbalanced STPHS.

3. Voltage supply modeling

Modeling three-phase voltage signals to be applied to the AC machine windings is the subject of the third section which exploits the structure shown in **Figure 3a** characterized by the relationship:

$$\sum_q i_q = 0 \tag{16}$$

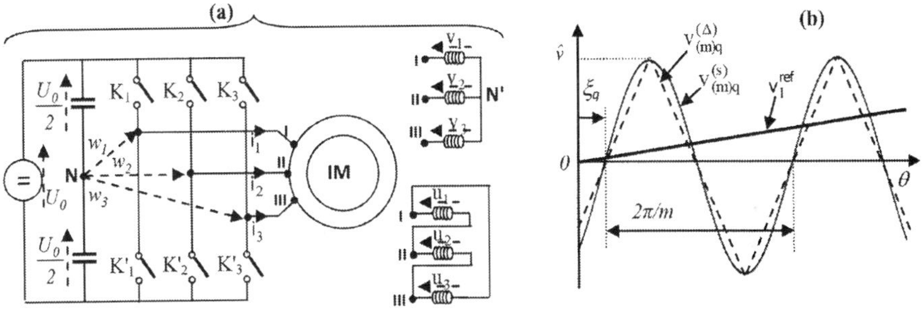

Figure 3. (a) IM connected to a three-phase PWM inverter, (b) control signals.

Two three-phase voltage systems appear in **Figure 3a**: that of v_q voltages and that of w_q voltages still qualified three-phase intermediate voltage system.

3.1. Characterization of the voltage STPHSs applied to the load

Let us consider an unbalanced w_q^k STPHS of rank "k". It can be assumed that this w_q^k system results from the sum of (C), (A), and (H) three-phase elementary systems:

$$w_q^k = w_q^{k_C} + w_q^{k_A} + w_q^{k_H} \tag{17}$$

- For star connected windings (**Figure 3a**), the voltages satisfy the following relationship:

$$w_q = v_q + v_{N'N} \tag{18}$$

which must also be valid pour each harmonic system.

- Consequently, for (C) and (A) systems, one can write:

$$\sum_q w_q^{k_{CorA}} = \sum_q v_q^{k_{CorA}} + 3v_{N'N}^{k_{CorA}} \tag{19}$$

As $\sum_q w_q^{k_{CorA}} = 0$, taking into account Eq. (16), leads to: $\sum_q v_q^{k_{CorA}} = 0$, so that $v_{N'N}^{k_{CorA}} = 0$. It results in the following equality: $v_q^{k_{CorA}} = w_q^{k_{CorA}}$.

- For (H) systems, $i_q^{k_H}$ currents cannot exist according to Eq. (16). It results that $v_q^{k_H} = 0$. As $w_q^{k_H} = w^{k_H}$ what may be "q," leads to the following equality: $w_q^{k_H} = v_{N'N}^{k_H}$. Thus, the potential of N' varies similarly as the (H) components present in the w_q system avoiding to impact the v_q system with these (H) components.

- For delta connected windings (**Figure 3a**), u_q is given par the relationship: $u_q = w_q - w_{q+1}$. It appears immediately that u_q system consists only in (C) and (A) three-phase harmonic systems.

These elementary considerations show that the stator windings, either they are star or delta connected, are not affected by the presence of (H) STPHSs present in the w_q three-phase system, whereas (C) and (A) STPHSs are fully reflected to the machine input terminals. That property has been used to lead the analysis done in the previous section about the force components. That is also this property which will be used to cancel some harmonic components existing in the machine supply. However, it is required to find easily the phase order sequence of the different w_q (or v_q) STPHSs. That is why a method is then proposed for this determination by considering the equivalent star connected load.

3.2. Modeling of the STPHSs

Therefore, an annoying force component generated by the switching can simply be removed by making (H) the w_q STPHS that is at its origin. It is at this level that the feature of the proposed strategy requires controlling the inverter using a three-phase carrier system. Let us denote v_q^{ref} and $v_{(m)q}^{(\Delta)}$ the phase "q" sinusoidal reference and triangular carrier signals, whose angular frequencies are, respectively, "ω" and mω (frequency $f_{PWM} = mf$). "m" is defined as the modulation index. To characterize analytically the w_q STPHS phase sequences, since triangular carriers leads to tedious calculations, the authors intend to make these determinations by substituting $v_{(m)q}^{(s)}$ sinusoidal carriers to $v_{(m)q}^{(\Delta)}$ triangular carriers, these two signals presenting the same peak values as shown in **Figure 3b**. Let us note that for a classical (SΔ) PWM, $v_{(m)q}^{(\Delta)}$ is the same whatever "q".

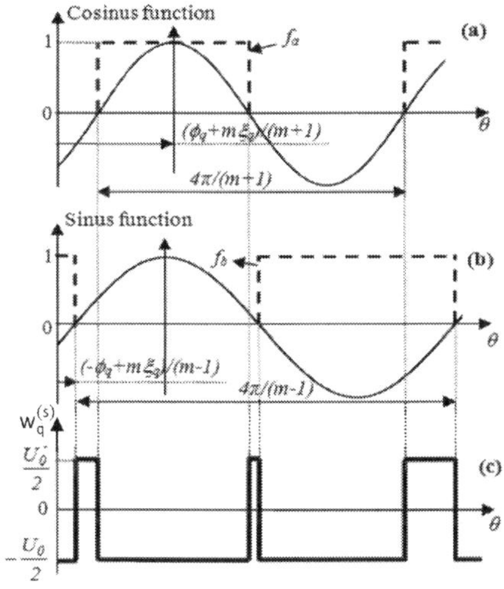

Figure 4. Principle of the $w_q^{(s)}$ determination.

Considering that the "r" adjusting coefficient is equal to the unit and a time referential tied to v_1^{ref}, introducing $\phi_q = (q-1)2\pi/3$, v_q^{ref} and $v_{(m)q}^{(s)}$ are expressed as:

$$\left.\begin{aligned} v_q^{ref} &= \hat{v}\sin\left(\theta-\phi_q\right) \\ v_{(m)q}^{(s)} &= \hat{v}\sin\left[m\left(\theta-\xi_q\right)\right] \end{aligned}\right\} \tag{20}$$

ξ_q is a phase "q" adjustable carrier phase difference.

- The control strategy is similar to that defined for a (SΔ) PWM: when $v_q^{ref} > v_{(m)q}^{(s)}$, the K_q switch is closed and $w_q^{(s)} = U_0/2$; for $v_q^{ref} < v_{(m)q}^{(s)}$, K'_q is closed and $w_q^{(s)} = -U_0/2$. The inequality $v_q^{ref} > v_{(m)q}^{(s)}$, taking into account Eq. (20) can be written as follows:

$$\cos\{[(m+1)\theta-\phi_q-m\xi_q]/2\}\sin\{[(m-1)\theta+\phi_q-m\xi_q]/2\}\leq 0 \tag{21}$$

Let us introduce the f_a and f_b logic functions defined as:

- $f_a = 1$ or 0 if the cosine is positive or negative, respectively.
- $f_b = 1$ or 0 if the sine is negative or positive, respectively.

It makes it possible to express the $w_q^{(s)}$ as:

$$w_q^{(s)} = U_0\left(2f_a f_b - f_a - f_b + 1/2\right) \quad (22)$$

Figure 4 shows the principle applied to determine $w_q^{(s)}$.

The f_a and f_b Fourier series can be written as:

$$\left.\begin{aligned} f_a &= \frac{1}{2} + \frac{2}{\pi}\sum_{n_1=0}^{\infty}\frac{(-1)^{n_1}}{2n_1+1}\cos\left\{\frac{2n_1+1}{2}\left[(m+1)\theta - \phi_q - m\xi_q\right]\right\} \\ f_b &= \frac{1}{2} + \frac{2}{\pi}\sum_{n_2=0}^{\infty}\frac{-1}{2n_2+1}\sin\left\{\frac{2n_2+1}{2}\left[(m-1)\theta + \phi_q - m\xi_q\right]\right\} \end{aligned}\right\} \quad (23)$$

where n_1 and n_2 are positive or null integers. $w_q^{(s)}$ can be expressed as:

$$w_q^{(s)} = \sum_{n_1=0}^{\infty}\sum_{n_2=0}^{\infty}\left(w_q^{(s)k_1} - w_q^{(s)k_2}\right) \quad (24)$$

where:

$$\left.\begin{aligned} w_q^{(s)k_1} &= \hat{w}^{(s)k_1}\sin\left(k_1\theta - N^-\phi_q - mN^+\xi_q\right) \\ w_q^{(s)k_2} &= \hat{w}^{(s)k_2}\sin\left(k_2\theta - N^+\phi_q - mN^-\xi_q\right) \end{aligned}\right\} \quad (25)$$

$$\left.\begin{aligned} N^+ &= 1 + n_1 + n_2 \\ N^- &= n_1 - n_2 \end{aligned}\right\} \quad (26)$$

$$\left.\begin{aligned} k_1 &= mN^+ + N^- \\ k_2 &= mN^- + N^+ \end{aligned}\right\} \quad (27)$$

Let us denote $W_0^{(s)} = 4U_0/\pi^2$. The component amplitudes that intervene in Eq. (25) are given by $(1)^{(1+n_1)}W_0^{(s)}/\left[(2n_1+1)(2n_2+1)\right]$. If k_1 is a positive integer, k_2 can be positive or negative.

http://dx.doi.org/10.5772/64520

To avoid use of negative frequencies but, also, of negative amplitudes, the relationships given by Eq. (25) will be rewritten as:

$$\left.\begin{array}{l} w_q^{(s)k_1} = \hat{w}^{(s)k_1} \sin\left(\left|k_1\right|\theta - N^-\phi_q - mN^+\xi_q - (1+n_1)\pi\right) \\ w_q^{(s)k_2} = \hat{w}^{(s)k_2} \sin\left(\left|k_2\right|\theta - N^+\phi_q - mN^-\xi_q - n_1\pi\right) \; for \; k_2 > 0 \\ w_q^{(s)k_2} = \hat{w}^{(s)k_2} \sin\left(\left|k_2\right|\theta + N^+\phi_q + mN^-\xi_q - (1+n_1)\pi\right) \; for \; k_2 < 0 \end{array}\right\} \tag{28}$$

where

$$\hat{w}^{(s)k_1 or k_2} = W_0^{(s)} / \left[\left(2n_1+1\right)\left(2n_2+1\right)\right] \tag{29}$$

k	$n_2\downarrow$ $n_1\rightarrow$	0	1	2	3	4	5
(a): $\lvert k_1\rvert$	0	m; (H); 1	2m + 1; (C); 3	3m + 2; (A); 5	4m + 3; (H); 7	5m + 4; (C); 9	6m + 5; (A); 11
	1	2m – 1; (A); 3	3m; (H); 9	4m + 1, (C); 15	5m + 2; (A); 21	6m + 3; (H); 27	7m + 4; (C); 33
	2	3m – 2; (C); 5	4m – 1; (A); 15	5m; (H); 25	6m + 1; (C); 35	7m + 2; (A); 45	8m + 3; (H); 55
	3	4m – 3; (H); 7	5m – 2; (C); 21	6m – 1; (A); 35	7m; (H); 49	8m + 1; (C); 63	9m + 2; (A); 77
	4	5m – 4; (A); 9	6m – 3; (H); 27	7m – 2; (C); 45	8m – 1; (A); 63	9m; (H); 81	10m + 1; (C); 91
	5	6m – 5; (C); 11	7m – 4; (A); 33	8m – 3; (H); 55	9m – 2; (C); 77	10m – 1; (A); 99	11m; (H); 121
(b): $\lvert k_1\rvert$	0	1; (C)	m + 2; (A)	2m + 3; (H)	3m + 4; (C)	4m + 5; (A)	5m + 6; (H)
	1	m – 2; (C)	3; (H)	m + 4; (C)	2m + 5; (A)	3m + 6; (H)	4m + 7; (C)
	2	2m – 3; (H)	m – 4; (A)	5; (A)	m + 6; (H)	2m + 7; (C)	3m + 8; (A)
	3	3m – 4; (A)	2m – 5; (C)	m – 6; (H)	7; (C)	m + 8; (A)	2m + 9; (H)
	4	4m – 5; (C)	3m – 6; (H)	2m – 7; (A)	m – 8; (C)	9; (H)	m + 10; (C)
	5	5m – 6; (H)	4m – 7; (A)	3m – 8; (C)	2m – 9; (H)	m – 10; (A)	11; (A)

Table 4. Characterization of $w_q^{(s)}$ STPHs generated by the (SS) PWM.

Table 4 presents, for the first n_1 and n_2 values, the characteristics of the STPHSs of ranks $\left|k_1\right|$ (a) and $\left|k_2\right|$ (b). Information given in **Table 4** concerns the rank of the harmonic and the phase sequence. Concerning $\left|k_1\right|$, the numerical values of $\left[\left(2n_1+1\right)\left(2n_2+1\right)\right]$ are also given. As $1/\left[\left(2n_1+1\right)\left(2n_2+1\right)\right]$ decreases very quickly when n_1 and n_2 increase, the analysis may cover only the first n_1 and n_2 values.

The main conclusion which can be extracted from **Table 4** is that a harmonic of rank "k" belongs only to the group k_1 or only to the group k_2. Furthermore, the rank of this harmonic appears only one time. As a result, considering the expression (28) to characterize the STPHSs, Eq. (24) can be rewritten as:

$$w_q^{(s)} = \sum_k w_q^{(s)k} = \sum_{k_1} w_q^{(s)k_1} - \sum_{k_2} w_q^{(s)k_2} \tag{30}$$

One can note first that these developments are valid whatever the (SS) PWM is, i.e., synchronous or asynchronous. Then, they show that, in this case, STPHSs are either balanced or Homopolar. Consequently, only (C) or (A) STPHSs constitute the $v_q^{(s)}$ three-phase system according to the study presented in Section 2.1. The fundamental value is obtained from the group k_2 for $n_1 = n_2 = 0$ and its amplitude is $W_0^{(s)}$.

It appears that this analytical model has considerable advantages because it makes it possible to determine the $w_q^{(s)}$ (or $v_q^{(s)}$) spectral contents and the harmonic phase sequences, without calculating the switching instants.

The problem now is to analyze under what conditions these results are applicable for a (SΔ) PWM and how the "r" value impacts on this property established for $r = 1$. The developments will be done considering a synchronous PWM.

3.3. PWM modeling validation

Let us point out that, from a practical point of view, the w_q three-phase system is not accessible. Therefore, the validation concerns the v_q spectral content for sinusoidal v_q^{ref}, $m = 55$, $f = 50Hz$, $U_0 = 520V$ and a single carrier wave, by comparing analytical, numerical, and experimental results for $r = 1$, then for $r \neq 1$.

All the experiments are made by connecting to the inverter outputs a three-phase, 50 Hz, 15 kW, two pole pair cage IM star connected operating at no load. A Focusrite sound card programmed with the C sound language makes it possible to control the switching generating single (or multiple) carrier(s) with sinusoidal (or triangular) shape(s). The v_q spectral content, both in amplitude and phase, is obtained using a spectrum analyzer. Numerical values come from FFT made with Matlab™ or using a computer program that exploits the calculated switching instants. Cross spectra of v_1^k and v_2^k give the phase differences for the main components ($\delta\hat{v}_q^k > 3\%$) so that they define their (C) or (A) phase sequences. Let us point out that $2\pi/3$ and $4\pi/3$ differences correspond, respectively, to a (C) and (A) systems.

The v_q spectral component relative amplitudes defined as percentage result from the relationship:

$$\delta\hat{v}_q^k = 100\,\hat{v}_q^k / \hat{v}_q^1 \tag{31}$$

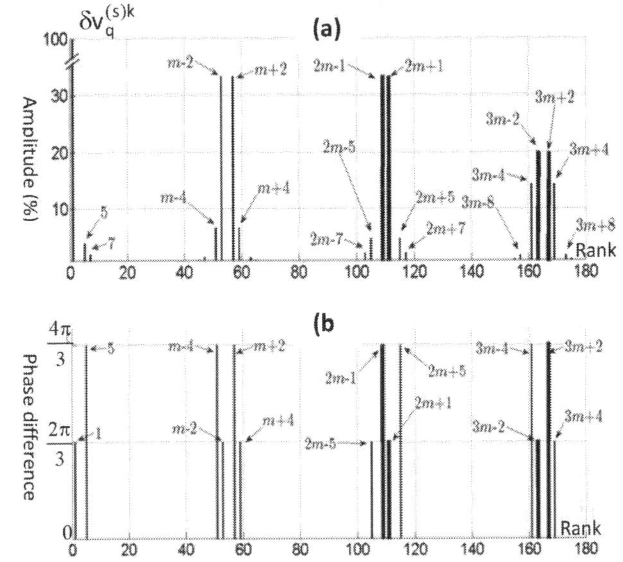

Figure 5. Experimental $v_q^{(s)}$ harmonic content for single sinusoidal carrier, m = 55, r = 1 and ξ_q = 0: (a) Harmonic content (b) phase differences.

3.3.1. *Single sinusoidal carrier*

- For r = 1, $w_q^{(s)}$ analytical harmonic ranks and phase sequences are given in **Table 4** for the first n_1 and n_2 values. As $\hat{w}_q^{(s)1} = W_0^{(s)}$ where $W_0^{(s)}$ = 210.75V, $\delta\hat{w}_q^{(s)k}$ values are obtained multiplying by one hundred the reversed numerical values given in **Table 4a**.

  The $v_q^{(s)}$ characteristics obtained experimentally for ξ_q = 0 are given in **Figure 5**, where the bold lines correspond to the terms resulting from the k_1 group (deduced from the analytical study). It appears that the number of components from the k_1 group is very limited and that these components appear only in the harmonic families centered on terms multiple of "m".

  Analytical and experimental results are in excellent correlation that validates the proposed methodology.

- The characteristics of the main STPHSs have been calculated for different ξ_q and r values.

- For $r = 1$, whatever the values of $\xi_{q'}$ the previous relative amplitudes are found. For the phases, the analytical and numerical values, for given value of $\xi_{q'}$ are almost the same, the gap increasing with "m". For example, for the $3m + 4$ rank component, the gap is inferior to 1° with the FFT and it is reduced when the switching instants are calculated.

- $r \neq 1$ leads to a similar spectral content than that is obtained for $r = 1$ with differences on the spectrum component amplitudes (the reference value in Eq. (31) is the fundamental component magnitude obtained experimentally for the considered "r" value). This analysis leads to a notable result which concerns the phases, and therefore the phase sequences, which are the same that for $r = 1 \ \forall r$.

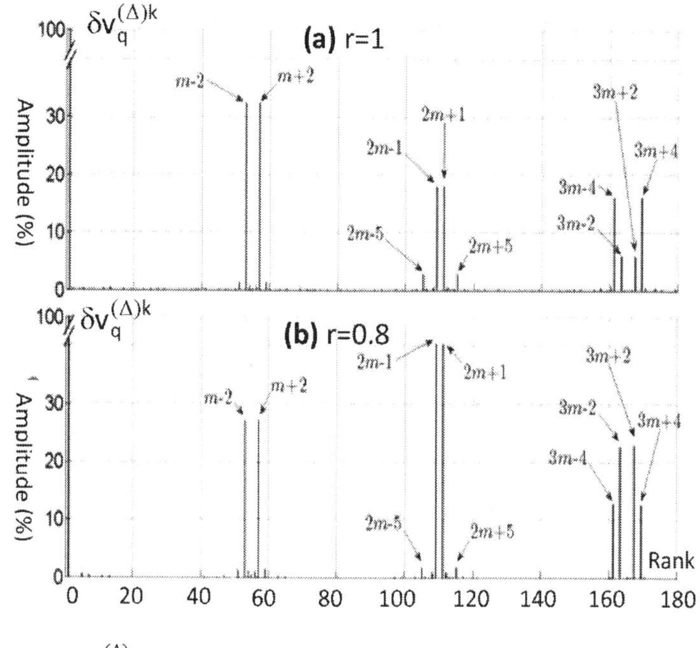

Figure 6. Experimental $v_q^{(\Delta)}$ harmonic content for single triangular carrier $m = 55$, and $\xi_q = 0$: (a) $r = 1$, (b) $r = 0.8$.

3.3.2. Single triangular carrier

Figure 6 shows the experimental $v_q^{(\Delta)}$ spectra obtained for r = 1 (**Figure 6a**) and $r = 0.8$ (**Figure 6b**). Eq. (31) makes it possible to calculate $\delta\widehat{w}_q^{(\Delta)k}$ taking into account that $\widehat{v}_q^{(\Delta)1} = rU_0/2$ (260 V for the considered U_0 value and r = 1). Comparing **Figures 5a** and **6a**, it appears, independently of the magnitude differences on the fundamental components according to the carrier type, that the same harmonic ranks are practically obtained. For components of low ranks (5 and 7), the magnitudes are notably amplified with the (SS) PWM, although these components do not intervene predominantly on the spectral content. One can also note that there is an inversion concerning the components of highest amplitude around $3m$.

The results obtained on the STPHS phase sequences highlight that they are independent not only from the carrier type but also from the "r" and ξ_q values. Consequently, they can be always deduced from Eq. (28). This property can be expressed as follows: the phase angles calculated with the expression established for a (SS) PWM, thus for $r = 1$, are also valuable for the harmonic components, which appears for a (SΔ) PWM and this, whatever the value of "r". The fact that "r" does not intervene appears in reference [27]. Compared to other methods [28], the one proposed by the authors allows simplifying the PWM analytical expressions which characterize its operating, thanks to the use of logical functions. For example, it has the advantage of not using a Fourier series with Bessel functions, as for the double-Fourier series methods used in reference [29].

The drawback of the proposed method is that it cannot be used if precise determination of the voltage harmonic component amplitudes is required. In fact, concerning this study, this drawback is not really a problem as it is in accordance with the manner of characterizing the force components.

This analysis allows concluding in general about the STPHSs. Indeed, whatever the method (analytical, modeling, or experimentation), the STPHSs satisfy, $\forall r$, the following equalities:

$$\left.\begin{array}{ll} v_q^k = w_q^k & for\ (C)\ and\ (A)\ systems \\ v_q^k = 0 & for (H)\ systems \end{array}\right\} \quad (32)$$

4. Simple voltage harmonic control

In this case, the method to simply remove a $v_q^{(s)}$ STPHS is presented denoting "x" or "y" the rank of the suppressed STPHS. As, in certain cases, that method leads to generate unbalances on the phases, the voltages will be characterized using "q". As a result, the new STPHSs will be characterized by the variables $w_{q(xory)}^{(s)k}$ or $v_{q(xory)}^{(s)k}$ that are sinusoidal functions, which will be defined to a constant that will be denoted, respectively, $C_{(x)}$ and $C_{(y)}$. To illustrate this method, first, two harmonics of the k_2 group, which contributes mainly on the definition of the spectral content, are concerned. Then, two harmonics of the k_1 group will be considered. From a practical point of view, all the results concern (SΔ) PWM with m = 55 and r = 1. The presentation of $v_q^{(\Delta)}$ spectra used to characterize the various components of a continuous line for (C) systems and a dotted line for (A) systems.

4.1. Suppression of k_2 group harmonics

4.1.1. Principle of the x = m + 2 or y = m − 2 rank harmonic elimination

- The case of the x=m+2 rank harmonic. As $k_2 > 0$ with $n_1 = 1$ and $n_2 = 0$, Eq. (28) leads to: $w_q^{(s)k_2} = \hat{w}^{(s)k_2}\sin\left[|k_2|\theta + \phi_q - m\xi_q - \pi\right]$. To make this system (*H*), the condition (33) must be satisfied:

$$m\xi_{q(x)} = \phi_q - C_{(x)} \tag{33}$$

Thus the modified systems, deduced from Eq. (28) can be expressed as:

$$\left.\begin{array}{l} w_{q(x)}^{(s)k_1} = \hat{w}^{(s)k_1}\sin\left[|k_1|\theta - (1+2n_1)\phi_q + N^{+}C_{(x)} - (1+n_1)\pi\right] \\ w_{q(x)}^{(s)k_2} = \hat{w}^{(s)k_2}\sin\left[|k_2|\theta - (1+2n_1)\phi_q + N^{-}C_{(x)} - n_1\pi\right] \; for\, k_2 > 0 \\ w_{q(x)}^{(s)k_2} = \hat{w}^{(s)k_2}\sin\left[|k_2|\theta + (1+2n_1)\phi_q - N^{-}C_{(x)} - (1+n_1)\pi\right] \; for\, k_2 < 0 \end{array}\right\} \tag{34}$$

The case of $y = m - 2$ rank harmonic. The method to suppress the v_{q}^{S}, $y = m - 2$ harmonic rank of the k_2 group with $k_2 < 0$ is similar to the previous one. This harmonic is defined by $n_1 = 0$ and $n_2 = 1$, Eq. (28) leads to $w_q^{(s)k_2} = \hat{w}^{(s)k_2}\sin\left[|k_2|\theta - \phi_q - m\xi_q - \pi\right]$. To make this harmonic system (H), it suffices to satisfy the equality in Eq. (35):

$$m\xi_{q(y)} = -\phi_q + C_{(y)} \tag{35}$$

That leads to characterize the modified system by Eq. (36):

$$\left.\begin{array}{l} w_{q(y)}^{(s)k_1} = \hat{w}^{(s)k_1}\sin\left[|k_1|\theta + (1+2n_2)\phi_q - N^{+}C_{(y)} - (1+n_1)\pi\right] \\ w_{q(y)}^{(s)k_2} = \hat{w}^{(s)k_2}\sin\left[|k_2|\theta - (1+2n_2)\phi_q - N^{-}C_{(y)} - n_1\pi\right] \; for\, k_2 > 0 \\ w_{q(y)}^{(s)k_2} = \hat{w}^{(s)k_2}\sin\left[|k_2|\theta + (1+2n_2)\phi_q + N^{-}C_{(y)} - (1+n_1)\pi\right] \; for\, k_2 < 0 \end{array}\right\} \tag{36}$$

- it can be observed that this control strategy does not modify the STPHS initial amplitudes and that all the $v_q^{(s)}$ modified STPHSs remain balanced according to the quantities which mutiply ϕ_q are integers. It appears also that the suppression of an harmonic is totally independent of the constant values so that the experiments are done for $C_{(x)} = C_{(y)} = 0$. On the other hand, Eqs 33 and 35 show that the method requires the use of three carriers.

4.1.2. Experimental measurements

Figure 7a presents the experimental spectrum for (SΔ) PWM with single carrier and r = 1. **Figure 7b** and **7c** show the spectra for controls corresponding to the suppression of the harmonics of rank $x = m + 2$ and $y = m - 2$ with, respectively, $\xi_{q(x)} = \phi_q/m$ and $\xi_{q(y)} = -\phi_q/m$. These spectra clearly show these deletions although the control laws result from mathematical developments which concern (SS) PWM.

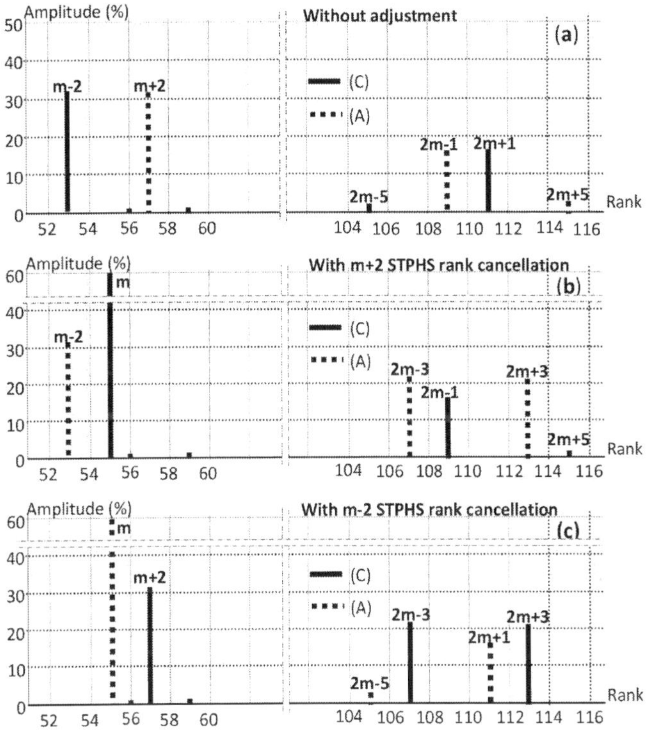

Figure 7. Relative experimental $v_q^{(\Delta)}$ spectra resulting from (b) $m + 2$ and (c) $m - 2$ rank harmonic cancelation for $r = 1$.

The suppression of $m + 2$ harmonic rank is accompanied by the disappearance of $2m + 1$ harmonic rank and the generation of m and $2m \pm 3$ harmonic ranks. Similarly, the suppression of $m - 2$ harmonic rank removes $2m - 1$ harmonic rank and generates harmonics of rank m and $2m \pm 3$. These particularities as well as the changes on the phase sequences which appear also on these figures, can easily be justified analytically using Eqs. 34 and 36. Let us note, in accordance with what has already been stated, that the theoretical relative amplitude of m rank component for (SS) PWM is 100% while, experimentally, for a (SΔ) PWM, it is only 60%.

4.2. Suppression of k_1 group harmonics

4.2.1. Principle of the x = 2m + 1 or y = 2m – 1 rank harmonic elimination

- The case of the x = 2m + 1 rank harmonic. $w_q^{(s)k_1}$ is defined by Eq. (28) with $n_1 = 1$ and $n_2 = 0$. That leads to $w_q^{(s)k_1} = \hat{w}^{(s)k_1}\sin\left[(2m+1)\theta - \phi_q - 2m\xi_q\right]$. To make this system (H), the following condition must be satisfied:

$$2m\xi_{q(x)} = -\phi_q + C_{(x)} \tag{37}$$

Thus the modified systems deduced from Eq. (28) can be expressed as:

$$\left.\begin{aligned}
w_{q(x)}^{(s)k_1} &= \hat{w}^{(s)k_1}\sin\left[|k_1|\theta + (1-n_1+3n_2)\phi_q/2 - N^+C_{(x)}/2 - (1+n_1)\pi\right]\\
w_{q(x)}^{(s)k_2} &= \hat{w}^{(s)k_2}\sin\left[|k_2|\theta - (2+n_1+3n_2)\phi_q/2 - N^-C_{(x)}/2 - n_1\pi\right]\ for\ k_2>0\\
w_{q(x)}^{(s)k_2} &= \hat{w}^{(s)k_2}\sin\left[|k_2|\theta + (2+n_1+3n_2)\phi_q/2 + N^-C_{(x)}/2 - (1+n_1)\pi\right]\ for\ k_2<0
\end{aligned}\right\} \tag{38}$$

- The case of the y = 2m – 1 rank harmonic. $w_q^{(s)k_1}$ is defined by Eq. (28) with $n_1 = 0$ and $n_2 = 1$ so that $w_q^{(s)k_1} = \hat{w}^{(s)k_1}\sin\left[(2m-1)\theta + \phi_q - 2m\xi_q - \pi\right]$. To make this system (H), the following equality must be satisfied:

$$2m\xi_{q(y)} = \phi_q - C_{(y)} \tag{39}$$

The modified systems, deduced from Eq. (28) can be written as:

$$\left.\begin{aligned}
w_{q(y)}^{(s)k_1} &= \hat{w}^{(s)k_1}\sin\left[|k_1|\theta - (1+3n_1-n_2)\phi_q/2 + N^+C_{(y)}/2 - (1+n_1)\pi\right]\\
w_{q(y)}^{(s)k_2} &= \hat{w}^{(s)k_2}\sin\left[|k_2|\theta - (2+3n_1+n_2)\phi_q/2 + N^-C_{(y)}/2 - n_1\pi\right]\ for\ k_2>0\\
w_{q(y)}^{(s)k_2} &= \hat{w}^{(s)k_2}\sin\left[|k_2|\theta + (2+3n_1+n_2)\phi_q/2 - N^-C_{(y)}/2 - (1+n_1)\pi\right]\ for\ k_2<0
\end{aligned}\right\} \tag{40}$$

http://dx.doi.org/10.5772/64520

- As previously:

  - the constants do not affect the results, so they will be assumed to be null

  - Equations 37 and 39 show that the method requires the use of three carriers

It can also be observed that this control strategy does not modify the STPHS initial amplitudes but that the quantities which multiply ϕ_q are not integers unless if the quantities between brackets are even. That means that many systems initially balanced become unbalanced. **Tables 5** and **6** show the modifications of $w_q^{(s)k}$ STPHS phase sequences, respectively, for 2m + 1 and 2m – 1 harmonic rank cancellations. HCA means that a system initially (C), (A), or (H) becomes unbalanced. Only the harmonic components which do not depend on m (k_2 group) and those around $6m$ (k_1 group), keep the same characteristics.

		(a) k_1 group				**(b) k_2 group**			
n_1		0	1	2	3	0	1	2	3
n_2	0	m (H)→(HCA)	2m + 1 (C)→(H)	3m + 2 (A)→(HCA)	4m + 3 (H)→(D)	1 (C)	m + 2 (A)→(HCA)	2m + 3 (H)→(A)	3m + 4 (C)→(HCA)
	1	2m – 1 (A)→(C)	3m (H)→(HCA)	4m + 1 (C)→(A)	5m + 2 (A)→(HCA)	–m + 2 (C)→(HCA)	3 (H)	m + 4 (C)→(HCA)	2m + 5 (A)→(C)
	2	3m – 2 (C)→(HCA)	4m – 1 (A)→(H)	5m (H)→(HCA)	6m + 1 (C)	–2m + 3 (H)→(A)	–m + 4 (A)→(HCA)	5 (A)	m + 6 (H)→(HCA)
	3	4m – 3 (H)→(C)	5m – 2 (C)→(HCA)	6m – 1 (A)	7m (H)→(HCA)	–3m + 4 (A) →(HCA)	–2m + 5 (C)→(H)	–m + 6 (H)→(HCA)	7 (C)

Table 5. Theoretical impact of the $2m + 1$ harmonic (group k_1) suppression on the $w_q^{(s)}$ spectral content.

		(a) k_1 group				**(b) k_2 group**			
n_1		0	1	2	3	0	1	2	3
n_2	0	m (H)→(HCA)	2m + 1 (C)→(A)	3m + 2 (A)→(HCA)	4m + 3 (H)→(A)	1 (C)	m + 2 (A)→(HCA)	2m + 3 (H)→(C)	3m + 4 (C)→(HCA)
	1	2m – 1 (A)→(H)	3m (H)→(HCA)	4m + 1 (C)→(H)	5m + 2 (A)→(HCA)	–m + 2 (C)→(HCA)	3 (H)	m + 4 (C)→(HCA)	2m + 5 (A)→(H)
	2	3m – 2 (C)→(HCA)	4m – 1 (A)→(C)	5m (H)→(HCA)	6m + 1 (C)	–2m + 3 (H)→(C)	–m + 4 (A)→(HCA)	5 (A)	m + 6 (H)→(HCA)
	3	4m – 3 (H)→(A)	5m – 2 (C)→(HCA)	6m – 1 (A)	7m (H)→(HCA)	–3m + 4 (A) →(HCA)	–2m + 5 (C)→(A)	–m + 6 (H)→(HCA)	7 (C)

Table 6. Theoretical impact of the $2m - 1$ harmonic (group k_1) suppression on the $w_q^{(s)}$ spectral content.

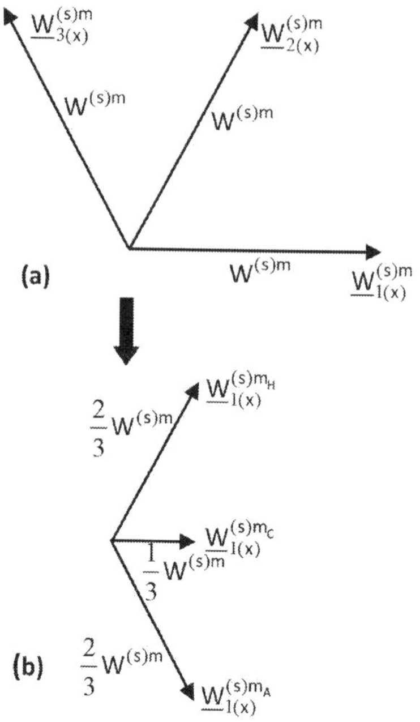

Figure 8. Cancellation of the harmonic x = 2*m* + 1: Components (C), (A) and (H) of *m* rank STPHS.

From magnetic noise point of view, it is necessary to characterize each STPHS with its (C), (A) and (H) components. They result from the following classical expressions which use time phasor variables and where $a = \exp(j2\pi/3)$:

$$\left.\begin{aligned} \underline{W}_{1(xory)}^{(s)k_C} &= (\underline{W}_{1(xory)}^{(s)k} + a\underline{W}_{2(xory)}^{(s)k} + a^2\underline{W}_{3(xory)}^{(s)k})/3 \\ \underline{W}_{1(xory)}^{(s)k_A} &= (\underline{W}_{1(xory)}^{(s)k} + a^2\underline{W}_{2(xory)}^{(s)k} + a\underline{W}_{3(xory)}^{(s)k})/3 \\ \underline{W}_{1(xory)}^{(s)k_H} &= (\underline{W}_{1(xory)}^{(s)k} + \underline{W}_{2(xory)}^{(s)k} + \underline{W}_{3(xory)}^{(s)k})/3 \end{aligned}\right\} \tag{41}$$

Let us consider, for example, the x = 2m + 1 harmonic rank suppression and the modified system given by Eq. (38). The "m" harmonic rank belongs to the k_1 group, and it is defined for $n_1 = n_2 = 0$ (**Table 5**). Considering Eq. (37) for $C_{(x)} = 0$, this "m" harmonic rank system can be expressed as:

$$w_{q(x)}^{(s)m} = \hat{w}^{(s)m} \sin\left(m\theta + (q-1)\frac{\pi}{3} - \pi \right) \tag{42}$$

http://dx.doi.org/10.5772/64520

It leads to **Figure 8(a)**, where the modulus of each time phasor variable is defined using its RMS value: $W^{(s)m} = \hat{w}^{(s)m}/\sqrt{2}$. Eq. (41) leads to the Phase 1 time phasor variable of each component given in **Figure 8(b)**. One can note that (H) and (A) systems have the same $2W^{(s)m}/3$ modulus, the (C) modulus is $W^{(s)m}/3$. As, theoretically, the relative amplitude of the m harmonic is equal to 1 (**Table 4**), one can deduced that the relative absolute values of the (C) and (A) systems of $v^{(s)m}$ are, respectively, equal to 33.3 and 66.6%.

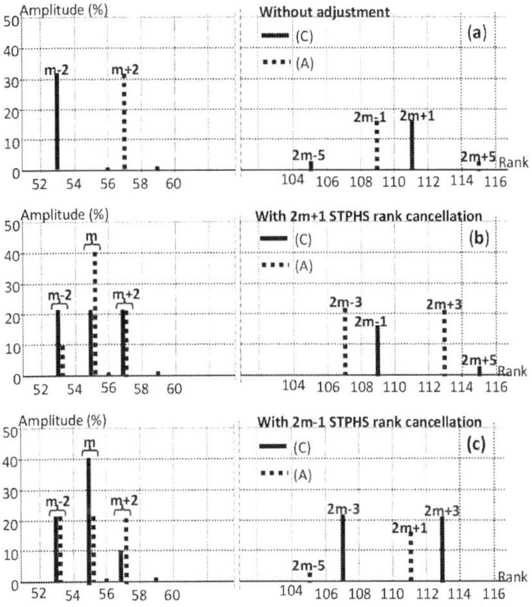

Figure 9. Relative experimental $v_q^{(\Delta)}$ spectra for $m = 55$ and $r = 1$ resulting from: (b) $2m + 1$ and (c) 2m – 1 rank harmonic cancelation.

4.2.2. Experimental measurements

- First, verifications about $v_q^{(\Delta)}$ make it possible to show, for the unbalanced harmonic systems, that, for a harmonic, the spectral components have the same amplitudes on the three phases. As the expressions deduced from the (SS) PWM lead to accurate values of the phase angles, the authors have used expression 41 to characterize the (C) and (A) amplitudes of the $v_q^{(\Delta)}$ systems. **Figures 9b** and **c** show the experimental $v_q^{(\Delta)}$ spectrum thus obtained for, respectively, 2m + 1 and 2m – 1 harmonic rank cancellation. This suppression leads:

- to new components of rank $2m \pm 3$ which are (C) or (A), as determined with the analytical model,

- to a unbalanced "m" rank component which is constitutes of (C) and (A) three-phase systems. The amplitudes of these latter present a sum always equal to 100%.

The validity of the proposed method is established as well as the theoretical approach proposed for characterizing the STPHS phase sequences. It is now possible to consider more complex control strategies in order, for example, to mitigate some of the negative effects caused by the removal of a single harmonic as the generation of new harmonic ranks in v_q spectral content. To do that, a procedure has been developed: the carrier phase jump which is presented in the following section.

5. Carrier-phase jump

To present this method, it will be shown how it is possible to minimize, for example, the effects of the reappearance of the harmonic system of rank "m" when the harmonic system of rank m + 2 is suppressed. The principle consists in associating with this harmonic of m + 2 rank, harmonic of m–2 rank, and alternately removing these harmonic systems. By exploiting the degree of freedom introduced by the constants makes it possible to optimize the minimization of harmonic of "m" rank.

5.1. Principle

The rank $x = m + 2$ harmonic is suppressed during the first v_q^{ref} period and the rank $y = m - 2$ harmonic during the second one. This transition is accompanied by transitions on all the modified STPHSs. Taking into account Eqs 34 and 36 leads to expressing a modified k rank STPHS using the general form:

$$w_{q(xory)}^{(s)k} = \hat{w}^{(s)k} \sin\left[k\theta - \psi_{q(xory)}^{(s)k} \right] \tag{43}$$

$\psi_{q(xory)}^{(s)k}$, defined by identification, depends on ϕ_q, $C_{(x)}$ and $C_{(y)}$. The Fourier series of resulting $w_{q(x,y)}^{(s)k}$ signal requires considering an interval that corresponds to $2T$. The two expressions of this signal on the considered 2 T interval are the following:

$$\left.\begin{aligned} w_{q(x)}^{(s)k} &= \hat{w}^{(s)k} \sin\left[2k(\theta - \psi_{q(x)}^{(s)k} / 2k) \right] \\ w_{q(y)}^{(s)k} &= \hat{w}^{(s)k} \sin\left[2k(\theta - \psi_{q(y)}^{(s)k} / 2k) \right] \end{aligned}\right\} \tag{44}$$

Let us denote h^k as $w_{q(x,y)}^{(s)k}$ rank harmonic. If $t = 0$ is the instant which corresponds to the passage of $w_{q(x)}^{(s)k}$ to $w_{q(y)}^{(s)k}$, it can be written as:

$$w_{q(x,y)}^{(s)k} = \sum_{h^k=1}^{\infty} \hat{w}_{q(x,y)}^{(s)k,h^k} \cos(h^k\theta - \beta_{q(x,y)}^{(s)k,h^k}) \tag{45}$$

where

$$\left.\begin{aligned} \hat{w}_{q(x,y)}^{(s)k,h^k} &= \hat{w}^{s(k)}\sqrt{(A_q^{k,h^k})^2 + (B_q^{k,h^k})^2} \\ tg\beta_{q(x,y)}^{(s)k,h^k} &= B_q^{k,h^k} / A_q^{k,h^k} \end{aligned}\right\} \tag{46}$$

Table 7 gives the expressions of A_q^{k,h^k} and B_q^{k,h^k} with $\psi_{q+}^{(s)k} = (\psi_{q(x)}^{(s)k} + \psi_{q(y)}^{(s)k})/2$, $\psi_{q-}^{(s)k} = (\psi_{q(x)}^{(s)k} - \psi_{q(y)}^{(s)k})/2$

	A_q^{k,h^k}	B_q^{k,h^k}	$\hat{w}_{q(x,y)}^{(s)k,h^k}$	$\beta_{q(x,y)}^{(s)k,h^k}$
h^k even $\neq 2k$	0	0	0	
h^k odd	$8k\dfrac{\sin\psi_{q+}^{(s)k}\sin\psi_{q-}^{(s)k}}{(2k+h^k)(2k-h^k)\pi}$	$-4h^k\dfrac{\cos\psi_{q+}^{(s)k}\sin\psi_{q-}^{(s)k}}{(2k+h^k)(2k-h^k)\pi}$		
	$\mp\dfrac{2}{\eta\pi}\sin\psi_{q+}^{(s)k}\sin\psi_{q-}^{(s)k}$	$\pm\dfrac{2}{\eta\pi}\cos\psi_{q+}^{(s)k}\sin\psi_{q-}^{(s)k}$	$\hat{w}^{(s)k}\dfrac{2}{\eta\pi}\left\|\sin\psi_{q-}^{(s)k}\right\|$	$\psi_{q+}^{(s)k} + \pi/2$
$h^k = 2k$	$-\sin\psi_{q+}^{(s)k}\cos\psi_{q-}^{(s)k}$	$\cos\psi_{q+}^{(s)k}\cos\psi_{q-}^{(s)k}$	$\hat{w}^{(s)k}\left\|\cos\psi_{q-}^{(s)k}\right\|$	$\psi_{q+}^{(s)k} + \pi/2$

Table 7. x↔y jump—Expressions of the Series Fourier coefficients for a k rank term.

On T, for given k, this method leads to associate with the initial harmonic components defined by Eq. (44):

- a component of Integer (I) rank "k" defined by the values given in **Table 7** for $h^k = 2k$,

- components defined for h^k odd, leading to harmonics of Non-Integer (NI) ranks.

The characteristics of these components are defined by the values given in the first row for h^k odd in **Table 7**.

As h^k odd can be expressed by $h^k = 2k \pm \eta$ with η odd, these NI rank harmonics are defined on "T' by $k \pm \eta/2$.

Insofar as these harmonics present significant values only for η first values, the second row for h^k odd gives the simplified expressions valid only under these conditions.

So the (I) rank "k" component is surrounded by NI rank harmonics.

Let us point out that for the fundamental component ($k = 1$), the expressions of $w_{q(x)}^{(s)1}$ and $w_{q(y)}^{(s)1}$ that can be deduced from Eqs 34 and 36 are the same. Thus, the carrier-phase jump has no effect on this component.

5.2. Numerical applications

The expressions of $\psi_{q(x)}^{(s)k}$ and $\psi_{q(y)}^{(s)k}$, grouped in **Table 8**, are obtained:

- by identification considering Eqs. (34) and (36), with k taking successively the values $m \pm 2$, m, $2m \pm 1$ and $2m \pm 3$,
- and by characterizing the quantities $\varepsilon_{q(xory)}$ with their general expressions given by Eqs. (33) and (35).

The $\psi_{q+}^{(s)k}$ and $\psi_{q-}^{(s)k}$ expressions, defined in functions of $C^+ = C_{(x)} + C_{(y)}$ and $C^- = C_{(x)} - C_{(y)}$, are also given in **Table 8**. These mathematical developments have a notable interest because the h^k rank harmonic amplitudes, which differ with the value of q, only depend on $\psi_{q-}^{(s)k}$ (see **Table 7**) and thus on C^+. The phases depends only on $\psi_{q+}^{(s)k}$ and thus on C^-.

Figure 10 shows the variations of $\left|\cos\psi_{q-}^{(s)k}\right|$ and $\left|\sin\psi_{q-}^{(s)k}\right|$ with C^+ for the previous "k" values and the three "q" values. In order to exploit these curves, it should be noted that for $k = 2m \pm 1$ and $2m \pm 3$, the horizontal scale must be divided by 2. The harmonic amplitudes are obtained by multiplying these quantities by the coefficients given in **Table 7**. As shown in **Figure 10**, for a given "q" value, the maximal value of the (I) rank "k" harmonic is associated with (NI) harmonics whose values are null.

The new STPHSs are all unbalanced, therefore, they can be defined by their (C), (A), and (H) components. The analysis made on these component moduli, and more especially on $\left|\cos\psi_{q-}^{(s)k}\right|$ and $\left|\sin\psi_{q-}^{(s)k}\right|$, for a given value of the (I) rank "k" component, shows that the coefficients which intervene in their definitions (multiplying the modulus given in **Table 7**) are independent from h^k, as well as the values given to C^+ and C^-. The numerical values of these coefficients are given also in **Table 8**.

	$\psi_{q(x)}^{(s)k}$	$\psi_{q(y)}^{(s)k}$	$\psi_{q+}^{(s)k}$	$\psi_{q-}^{(s)k}$	(C)	(A)	(H)
$m-2$	$-\phi_q - C_{(x)} + \pi$	$C_{(y)} + \pi$	$\pi - (\phi_q + C^-)/2$	$-(\phi_q + C^+)/2$	0	0.5	0.5
$m+2$	$-C_{(x)} + \pi$	$\phi_q + C_{(y)} + \pi$	$\pi + (\phi_q - C^-)/2$	$-(\phi_q + C^+)/2$	0.5	0	0.5
m	$\phi_q - C_{(x)} + \pi$	$-\phi_q + C_{(y)} + \pi$	$\pi - C^-$	$\phi_q - C^+/2$	0.5	0.5	0
$2m-1$	$\phi_q - 2C_{(x)} + \pi$	$2C_{(y)} + \pi$	$\pi - C^- + \phi_q/2$	$-C^+ + \phi_q/2$	0.5	0	0.5
$2m+1$	$-2C_{(x)}$	$-\phi_q + 2C_{(y)}$	$\pi - C^- + \phi_q/2$	$-C^+ + \phi_q/2$	0	0.5	0.5
$2m-3$	$-\phi_q - 2C_{(x)} + \pi$	$\phi_q + 2C_{(y)} + \pi$	$\pi - C^-$	$-\phi_q - C^+$	0.5	0.5	0
$2m+3$	$-\phi_q - 2C_{(x)}$	$\phi_q + 2C_{(y)}$	$-C^-$	$-\phi_q - C^+$	0.5	0.5	0

Table 8. Expressions of $\psi_{q(x)}^{(s)k}$ and $\psi_{q(y)}^{(s)k}$ and (C), (A) and (H) component characterization.

First, one can note that a carrier-phase jump which involves in the transition an (*H*) STPHS, leads to corresponding $w_q^{(s)}$ STPHS which contains also (*H*) component. In this case, the $v_q^{(s)}$ STPHSs are balanced as shown in **Table 8**. For all the other cases, the $v_q^{(s)}$ STPHSs are unbalanced.

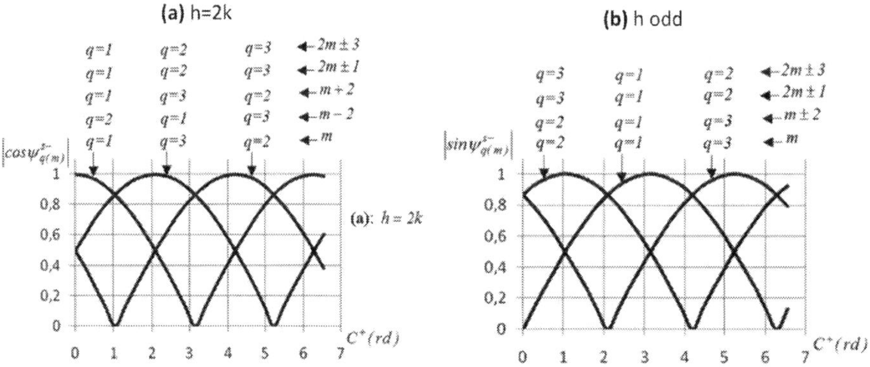

Figure 10. Variations of $\left|\cos\psi_{q(m)}^{s-}\right|$ and $\left|\sin\psi_{q(m)}^{s-}\right|$ with C^+.

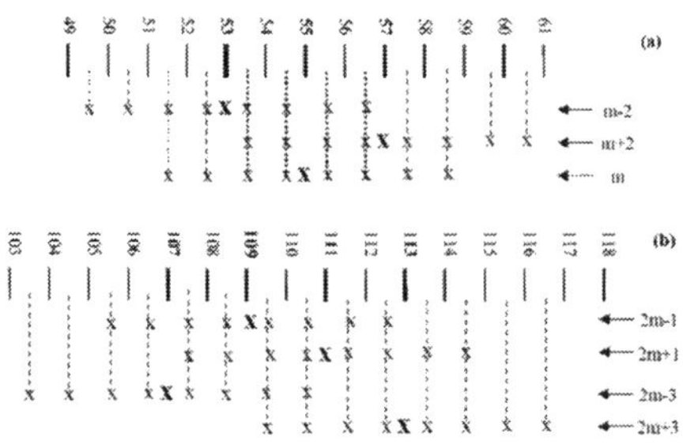

Figure 11. NI rank harmonics: (a) around m; (b) around $2m$.

The NI rank harmonics are subject to the second remark. They form balanced STPHSs for the $m \pm 2$ and $2m \pm 1$ (I) rank harmonics, and unbalanced STPHSs for all the other considered (I) rank harmonics. These harmonics combine together as it appears in **Figure 11** which considers the quantities that appear around m (**Figure 11a**) and around $2m$ (**Figure 11b**). For this study, the number of NI rank harmonics considered is limited to the η first four values. It results that the characteristic predetermination of NI rank harmonics needs to add several terms taking into account the $\beta_{q(x,y)}^{(s)k,h^k}$ quantities given in **Table 7**, which depend on $\psi_{q+}^{(s)k}$, and consequently on C^- quantities as it appears in **Table 8**.

5.3. Experimental validation for (SS) PWM

The aim of these applications is to validate the theoretical developments corresponding to the previous case ($m = 55$, $r = 1$, $x = m + 2$ and $y = m - 2$).

• First, the constants are such as $C_{(x)} = C_{(y)} = 0$, defining $C^+ = 0$. The analytical developments lead to the $v_{q(x,y)}^{(s)}$ component relative amplitudes in % given in **Table 9**. The quantities for NI rank harmonics are calculated for $\eta = 1$. Let us point out that only two lines, $q = 1$ and $q = 2$, constitute **Table 9** because, for this case, the $q = 2$ and 3 components present identical amplitudes. When the numerical values for $q = 1$ and 2 are the same, the STPHSs are balanced.

	$m(100\%)$		$m \pm 2(33.3\%)$		$2m \pm 1(33.3\%)$		$2m \pm 3(20\%)$	
	I	NI	I	NI	I	NI	I	NI
$q = 1$	100	0	16.65	10.64	16.65	10.64	20	0
$q = 2$	50	55	16.65	10.64	16.65	10.64	10	11.1

Table 9. $v_{q(x,y)}^{(s)}$ component relative amplitudes for $C^+ = 0$.

The $v_{q(x,y)}^{(s)}$ spectra corresponding to this (SS) PWM are given in **Figure 12**. Let us consider the (I) rank harmonics. One can note a very good correspondence between the predetermined values and those deduced from the spectra.

Concerning the NI rank harmonics, let us consider the $m + 0.5$ rank harmonic which results from the combination of three terms as it appears in **Figure 11a**:

- the harmonic related to $k = m$ for $\eta = 1$, whose relative amplitude is 0% for $q = 1$ and 55% for $q = 2$ and 3,

- the harmonic related to $k = m + 2$ for $\eta = 2$, whose relative amplitude is 5.32% for the three phases,

- the harmonic related to $k = m - 2$ for $\eta = 3$, whose relative amplitude is 3.55% for the three phases.

The approach is identical to characterize $m - 0.5$ rank harmonic. One can estimate that the components of $m \pm 0.5$ ranks have an amplitude of a few percents for $q = 1$ and an amplitude of the order of 50–60% for $q = 2$ and 3. These quantities correspond to those measured on the spectra. A similar approach for the other NI harmonics leads to similar conclusions.

- The constants are adjusted such as $C_{(x)} = 0$ et $C_{(y)} = \pi$, leading to $C^{+} = \pi$. **Table 10** presents the $v_{q(x,y)}^{(s)}$ component relative amplitudes given in % deduced from the analytical developments. Only the harmonics around "m" rank are considered because taking into account the periodicity of curves presented in **Figure 10**, the spectral content around $2m$ is the same than the content obtained for $C^{+} = 0$. The spectral contents presented in **Figure 13**, which concern only harmonics around "m", show the good correspondence between predetermined and measured values that validates the analytical relationships.

	$m(100\%)$		$m \pm 2(33.3\%)$	
	I	NI	I	NI
$q = 1$	0	64	16.65	10.64
$q = 2$	86.6	32	16.65	10.64

Table 10. $v_{q(x,y)}^{(s)}$ component relative amplitudes for $C^{+} = \pi$.

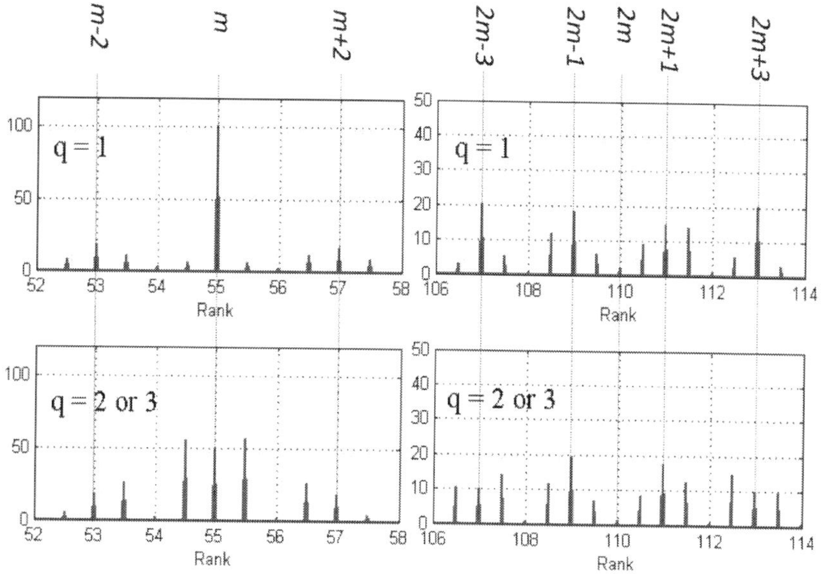

Figure 12. (SS) PWM, $v_{1and2}^{(s)}$ spectra around m and $2m$ for $C^{+} = 0$.

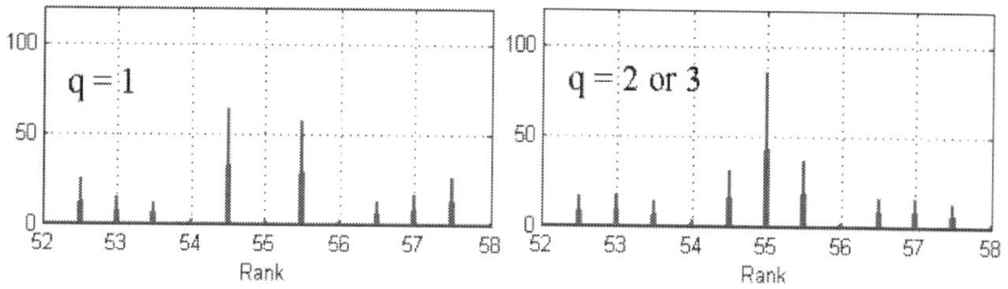

Figure 13. (SS) PWM, $v_{1and2}^{(s)}$ spectra around m for $C^{+} = \pi$.

5.4. Experimental validation for (SΔ) PWM

The spectra measured for a (SΔ) PWM, given in **Figures 14** and **15**, have the same properties when the constants are changed; only the amplitudes are modified. For example, the spectra shown in **Figures 14** and **15** correspond to the spectra given in **Figures 12** and **13** obtained for a (SS) PWM.

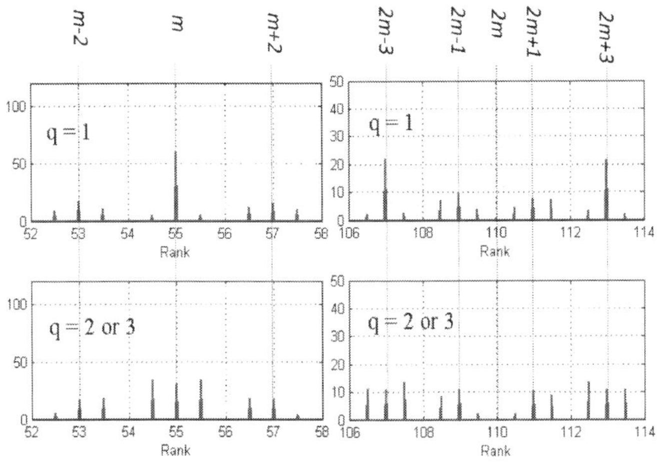

Figure 14. (SΔ) PWM, $v_{1and2}^{(\Delta)}$ spectra around m and $2m$ for $C^{+} = 0$.

The validity of developments made on a (SS) PWM to characterize the spectral content of a (SΔ) PWM is again demonstrated on this quite particular control technique which results, compared with the original spectrum shown in **Figure 6a**, to a spread spectrum with a significant changes of the amplitudes of some initial harmonic components.

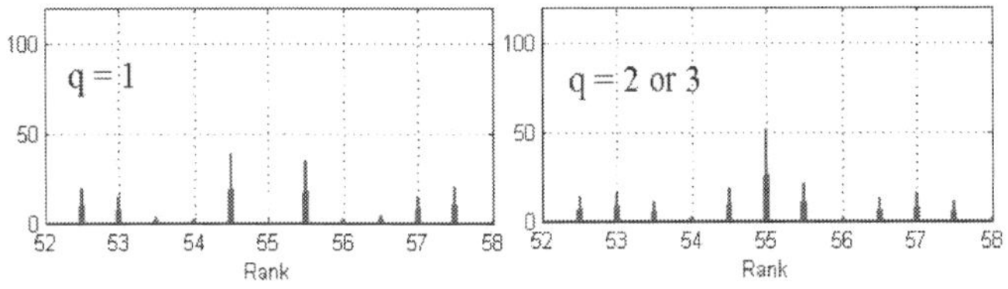

Figure 15. (SΔ) PWM, $v_{1and2}^{(\Delta)}$ spectra around m for $C^{+} = \pi$.

6. Use of carriers of different frequencies

6.1. Principle

The principle is shown considering the theoretical model of a (SS) PWM. The modulation index m_q concerns the phase "q". It will be supposed that m_q differs with "q" satisfying, for example, the equalities $m_{q-1} = m - \Delta m$, $m_{q+1} = m + \Delta m$. **Table 4** which gives the $w_q^{(s)}$ harmonic content shows that

- the harmonic of ranks 1, 3, 5, 7, ..., are independent of the values given to m_q and, thus, the fundamental is not modified when carrier frequencies are different
- the other harmonics define groups of components centered:

$$\left.\begin{array}{ll} \text{-on } m_q: & m_q, m_q \pm 2, m_q \pm 4, m_q \pm 6, \\ \text{-on } 3m_q: & 3m_q, 3m_q \pm 2, 3m_q \pm 4, 3m_q \pm 6, \\ \text{-on } 5m_q: & 5m_q, 5m_q \pm 2, 5m_q \pm 4, 5m_q \pm 6, \\ \text{-on } 2m_q: & 2m_q \pm 1, 2m_q \pm 3, 2m_q \pm 5, \\ \text{-on } 4m_q: & 4m_q \pm 1, 4m_q \pm 3, 4m_q \pm 5, \\ \text{-on } 6m_q: & 6m_q \pm 1, 6m_q \pm 3, 2m_q \pm 5, \end{array}\right\} \qquad (47)$$

The initial structure in **Figure 3a** can also be presented using the scheme given in **Figure 16** where a harmonic, function of m_q, must also be distinguished according to the considered phase "q". It results that an harmonic rank will be denoted using two superscripts $k^{q,j}$.

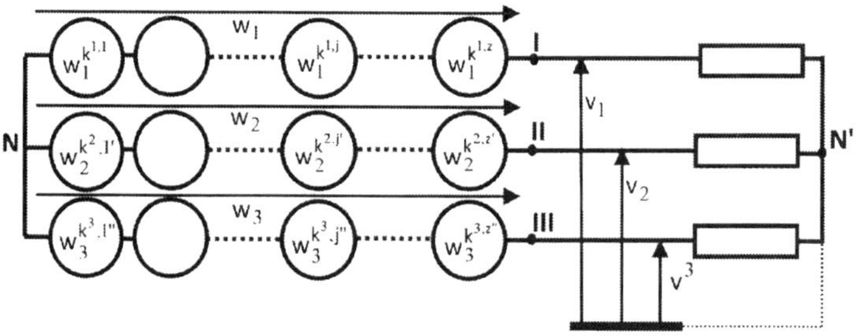

Figure 16. Equivalent structure of the converter associated to the IM.

To analyze how an harmonic of rank "k" modifies the behavior of the load, the superposition principle is applied and the generator giving a signal of frequency kf is searched for each phase. Then, after identification of these different generators, all the others are switched off. Let us underline that this approach concerns all the harmonics different from the components of rank 1, 3, 5, 7 ...., for the latter, the voltage STPHSs applied to the load will be balanced because the (H) component cannot exist at the terminal connections of the load.

It results that an harmonic of "k" rank can, for example, concerning the Phase 1, belong to the group of components centered on m_1 and, concerning the Phase 2, to the group of component centered on $2m_2$. This example shows that three cases have to be considered:

- *Case 1.* If a STPHS of rank "k," which exists on a phase, is also on the two other phases, then the structure in **Figure 17a** is obtained where the three generators constitute an imbalanced three-phase system where $k^{1,j} = k^{2,j'} = k^{3,j''} = k$.

- *Case 2*. If a STPHS of rank "k," which exists on Phase 1, is only on one of the two other phases (Phase 2, for example), then the structure in **Figure 17b** is obtained where $k^{1,j} = k^{2,j'} = k$ and where the RMS voltage generators have, a priori, no link.

- *Case 3*. If a STPHS of rank "k," which exists on Phase 1, is not on the two other phases, then the structure in **Figure 17c** has to be considered where $k^{1,j} = k$.

- The Case 1 has already been considered: the intermediate system has to be decomposed in (C), (A), and (H) components using Eq. (41) and considering that only the (C) and (A) systems are applied to the load.

- For the Case 2 it appears, considering the example given in **Figure 17b**, that $\underline{V}^k_1 = (2\underline{W}^k_1 - \underline{W}^k_2)/3$, $\underline{V}^k_2 = (2\underline{W}^k_2 - \underline{W}^k_1)/3$, $\underline{V}^k_3 = -(\underline{W}^k_1 + \underline{W}^k_2)/3$. This STPHS that appears at the load inputs is unbalanced without (H) component.

- Concerning the Case 3, **Figure 17c** leads to $\underline{V}^k_1 = 2\underline{W}^k_1/3$, $\underline{V}^k_1 = -\underline{W}^k_1/3$. As previously, this STPHS that appears at the load inputs is unbalanced without (H) component.

- When the structures of circuits in **Figures 17b** and **17c** are changed (by considering, for example, that the generator of **Figure 17c** that supplies the circuit is not connected to Phase 1 but to Phase 2, the previous equations can still be used but need an adaptation of the indexes.

Figure 17. Different configurations of the equivalent circuit with *k*.

6.2. Numerical application—experimental validation

The case of $m_1 = 45$, $m_2 = 55$, and $m_3 = 65$ for f = 50 Hz and r = 1 is considered.

6.2.1. (SS) PWM

Taking into account Eq. (47), it appears immediately that all the harmonics have odd ranks. Considering only the two first groups of components, the theoretical spectra given in **Figure 18** show that:

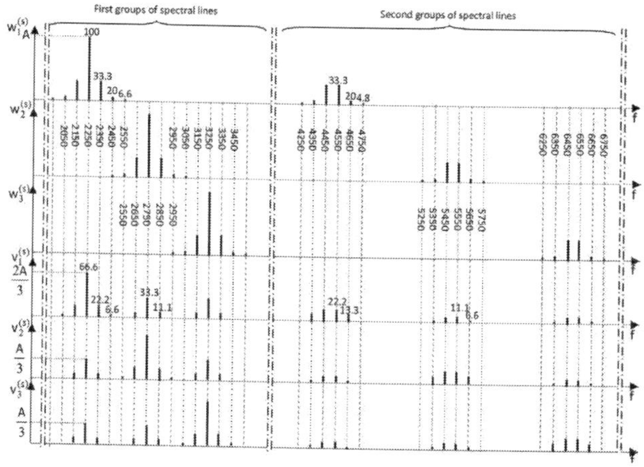

Figure 18. Principle of frequency shift carriers for (SS) PWM, f = 50 Hz and m_1 = 45, m_2 = 55 et m_3 = 65.

- For the first group of components, the equivalent circuits are the same as the circuit given in **Figure 17c**. The component m_1f (2250 Hz) of $w_1^{(s)}$, whose theoretical amplitude is 100% (**Table 4**), is dispatched on $v_q^{(s)}$ with an amplitude of 66.6% for q = 1 and 33.3% for q = 2 and 3. Concerning $w_q^{(s)}$, there is, for example, coincidences of the component $(m_1 + 4)f$ of $w_1^{(s)}$ with the component $(m_2\text{-}6)f$ of $w_2^{(s)}$ (2450 Hz). However, the amplitudes of these theoretical components are, respectively, 6.6 and 2.8% so that the resulting effect to be analyzed with the scheme given in **Figure 17b** will nearly not appear in the $v_q^{(s)}$ spectra. Consequently, this interaction is ignored.

Coincidences exist, in general, for all the components, but if the quantities m_q are chosen correctly, they involve two generators on the three that have almost zero voltages so that one has only to consider the configuration given in **Figure 17c**.

- The previous remark can be generalized for all the second groups of components shown in **Figure 18**. It must be noted that the third group of component (not shown) modifies significantly the amplitudes of the components between 6250 and 6750 Hz. Indeed, if the third group is considered for q = 1, the components which will interfere with the existing components have, for the most important of them, the following characteristics:

- $3m_1f$ = 6750 Hz whose theoretical amplitude is 11.1%,

- $(3m_1\text{-}2)f$ = 6650 Hz whose theoretical amplitude is 20%

- $(3m_1\text{-}4)f$ = 6550 Hz whose theoretical amplitude is 14.3%

- $(3m_1\text{-}6)f$ = 6450 Hz whose theoretical amplitude is 3.7%

In this case, the analysis has to exploit the scheme in **Figure 17b**. It results that what appears in Phase 1 is different from what can be observed on Phases 2 and 3 which are not modified by the third group of components.

6.2.2. *(SΔ) PWM*

The experimental $v_2^{(\Delta)}$ and $v_3^{(\Delta)}$ spectra (without decomposition considering (C) and (A) systems) for a (SΔ) PWM for the conditions corresponding to the theoretical study, are shown in **Figure 19**. The spectral distribution in **Figure 18** is verified, except for the amplitudes.

This technic which consists in using three carriers of different frequencies leads, if comparing to a classical (SΔ) PWM, to a splintering of the spectrum with a decrease of the initial components.

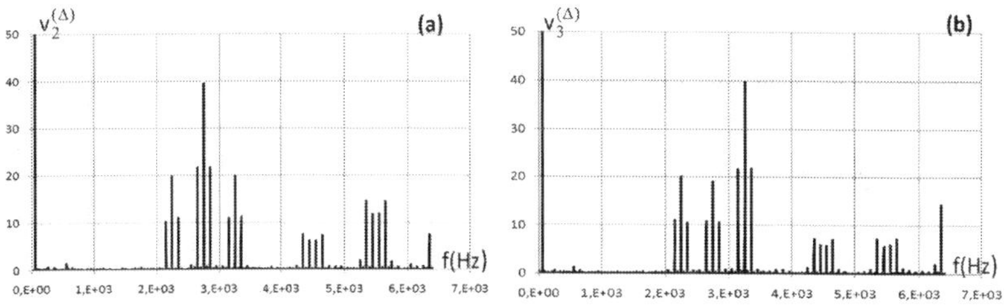

Figure 19. Experimental validation of the principle using carriers of different frequencies for f = 50 Hz and m_1 = 45, m_2 = 55 and m_3 = 65.

Concerning this presentation, it seems difficult to develop further as the number of combinations on the choice of values to assign to the mq quantities is infinite justifying that, from an analytical point of view, only the principle is presented.

7. Experiments on noise and vibrations

Certain machines have been redesigned in order to be more efficient when they operate at variable speed using PWM inverters. Precautions have also been taken to minimize noise problems. Unfortunately, for some of them, severe acoustic problems due to the STPHSs appear when they operate. It is in this context that the presented methods were developed. For confidential reasons, it is not possible to give the results about these specific experiments. Thus the presented results concern 2 IMs available in our laboratory with the following characteristics:

M1: cage IM—15 kW, 50 Hz, p = 2, 1456 rpm, 380 V, 24.8A, N^s = 48, N^r = 36.

M2: wound rotor IM—15 kW, 50 Hz, p = 3, 950 rpm, 365 V, 28A, N^s = 72, N^r = 54.

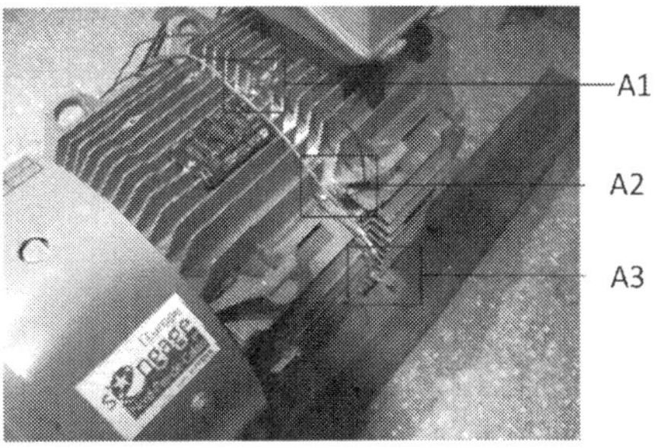

Figure 20. Positions of the accelerometers on M1.

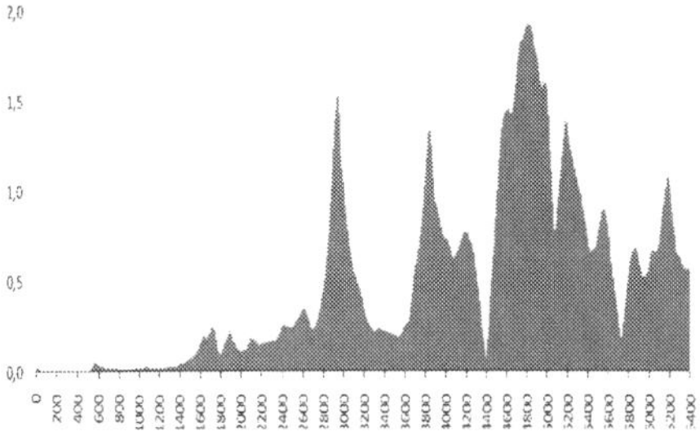

Figure 21. Frequency response of the M1 outer casing.

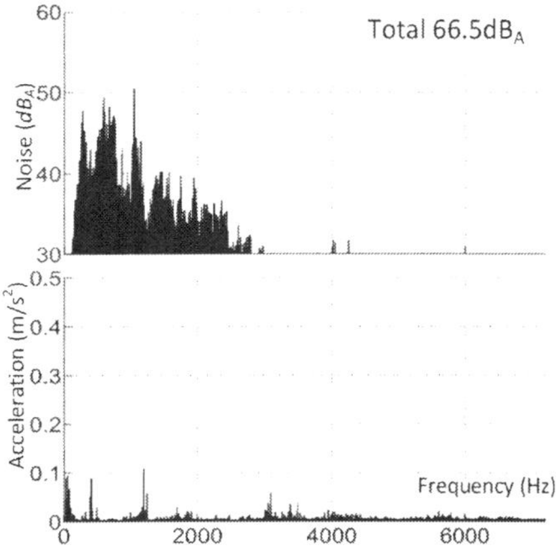

Figure 22. Noise and vibrations of M1 connected to the grid.

http://dx.doi.org/10.5772/64520

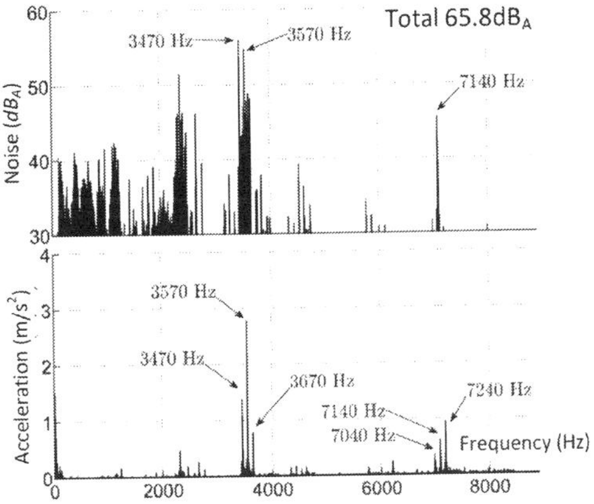

Figure 23. Noise and vibrations of M2 connected to the grid.

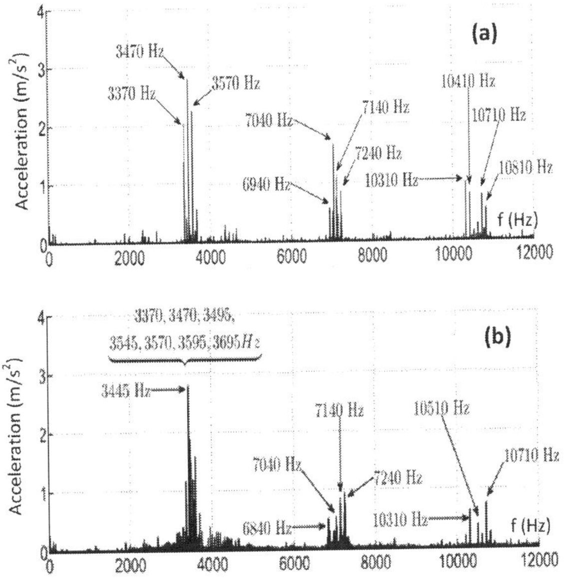

Figure 24. Vibrations of M2:

- (a) connected to a single carrier (SΔ) PWM—m =70.4, r = 1.

- (b) connected to (SΔ) PWM—m=70.4, r = 1 and carrier phase jump.

Alternative cancellation of $2m + 1$ and $2m–1$ harmonic ranks.

Since the control techniques must be adapted to the problems that appear on machines taking into account their design, the presented results concern either M1 or M2. Note that these two IMs are not especially sensitive to magnetic noise generated by STPHSs and, therefore, will not lead, in terms of reducing vibrations and noise, to performance that can described as exceptional. However, the effects are significant enough to validate our developments.

- **Figure 20** shows M_1 equipped with three accelerometers A1, A2, and A3 (A2 and A3 are shifted spatially from 45° to 90° with respect to A1) to, on the one hand, characterize the mode numbers of the Maxwell force components and, on the other hand, to be sure to consider the effects generated by unbalanced STPHSs in accordance with what was presented in section II.3. **Figure 21** gives the results of nodal analysis performed on M_1 that shows that it is particularly sensitive to frequencies around 2800 Hz, 3800 Hz, 4800 Hz, and 6200 Hz.

  For the test results to be presented, the IMs are supplied by fundamental 50 Hz three-phase systems with an implementation of a (SΔ) PWM for $r = 1$.

- **Figures 22** and **23** show the noise and vibration spectra of, respectively, M1 and M2 directly connected to the grid. In this case, in terms of noise, the mechanical, aerodynamic and magnetic effects appear simultaneously. As can be noted, these "low frequency" effects generate a relatively significant overall noise (66.5 dBA for M1 and 65.8 dBA for M2). The M2 spectra shows more clearly the slotting effect by covering a wider frequency range given the highest values of N^s and N^r. In this case, the slotting effect will interfere with the noise components generated by the STPHSs as seen in **Figure 24** where $m = 70.4$. Note that specific procedures have been developed to minimize the slotting magnetic noise as presented in reference [30].

- **Figure 24a** shows the M2 acceleration spectra using a conventional (SΔ) PWM inverter that uses only one carrier. Note that the overall noise is 65.8 dBA when M2 is supplied directly by the grid and 66.4 dBA when M2 is supplied via the inverter. As already mentioned, this machine is not particularly sensitive to STPHSs though, from a vibrational point of view, the impact on first and second groups of spectral lines, respectively, centered on 3520 and 7040 Hz is clearly visible.

  The carrier-phase jump has been implemented on M2 to reduce the effects of STPHSs centered on 2m eliminating alternatively harmonics of ranks $2m + 1$ and $2m - 1$. To optimize this control, analytical developments show that the constant values must be adjusted as follows: $C_{(x)} = 0$, $C_{(y)} = 8\pi/3$. **Figure 24b** shows the impact of this control strategy on the spectral content of vibrations in areas that contain these harmonics. Spread spectrum with reduced amplitudes of the initial components is clearly visible with the appearance of an impact on the third group of components.

- **Figure 25** shows in terms of M1 vibrations, the impact of an adaptation of carrier frequencies with $m_1 = 39$, $m_2 = 48$, and $m_3 = 57$. Again, the positive impact of this control strategy is clear.

http://dx.doi.org/10.5772/64520

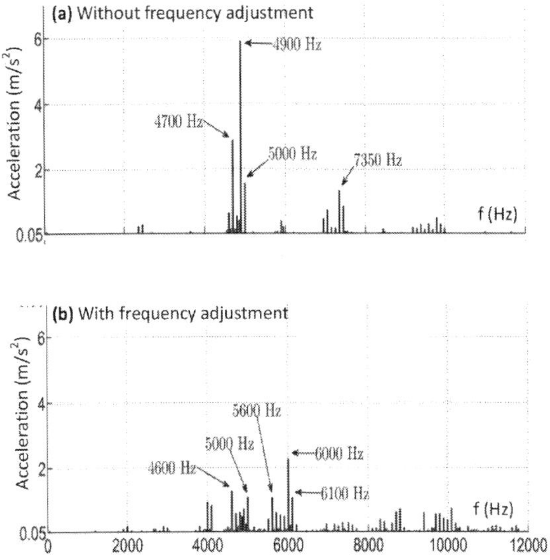

Figure 25. Vibrations of M1—Impact of a carrier frequency shift.

8. Conclusion

The proposed calculation method is particularly interesting, because of its implementation simplicity, to characterize the STPHS phase sequences that have a direct impact on the definition of the force components at the origin of the magnetic noise generated by AC machines. On the other hand, it allows anticipating easily over the entire audible frequency range of the spectrum, the consequences of certain control strategies for reducing this noise. Regarding the strategies proposed, they have been analyzed on the basis of experiments on noise and radial vibrations.

Author details

Jean-François Brudny*, Fabrice Morganti and Jean-Philippe Lecointe

*Address all correspondence to: jfrancois.brudny@univ-artois.fr

Laboratoire Systèmes Electrotechniques et Environnement (LSEE), Univ. Artois, Béthune, France

References

[1] Murphy JMD, Turnbull FG. Power electronic control of AC motors, Pergamon Press, New York, 1988.

[2] Romary R, Brudny JF. Harmonic torques of electrical AC electrical drives. Industrial Electronics Handbook. 2nd ed., CRC Press & IEEE Press. Part II, Chapter 11, février 2011. ISBN: 9781439802854, ISBN 10: 1439802858. p. 10.1-10.27.

[3] Khomfoi S, Kinnares V., Viriya P. Influence of PWM characteristics on the core losses due to harmonic voltages in PWM fed induction motors. In Proceedings of Power. Eng. Society Winter Meeting. Publisher IEEE, DOI: 10.1109/PESW2000.849991, Vol. 1, 2000, pp. 365–369.

[4] Boglietti A, Cavagnino A. Iron loss prediction with PWM supply: An overview of proposed methods from an engineering application point of view. Electric Power Systems Research 80, N 9, 2010, pp. 1121–1127.

[5] Iida S, Okuma Y, Masukawa S, Miyairi S, Bose BK. Study on magnetic noise caused by harmonics in output voltages of PWM inverter. IEEE Trans. on Industrial Electronics, 38, N 3, 1991, pp. 180–186.

[6] Pellerey P, Favennec G, Lanfranchi V, Friedrich G. Active reduction of electrical machines magnetic noise by the control of low frequency current harmonics. In: Proceedings of IECON 2012—38th Annual Conference on IEEE Industrial Electronics Society, October 2012, pp. 1654–1659.

[7] Mihaila V, Duchesne S, Roger D. A simulation method to predict the turn-to-turn voltage spikes in a PWM fed motor winding. IEEE Trans. on Dielectric and Electrical Insulation, Vol. 18, N 5, 2011, pp. 1609–1615.

[8] Cavallini A, Fabiani D, Montanari GC. Power electronics and electrical insulation systems—Part 2: life modeling for insulation design. DIES, IEEE Electrical Insulation Magazine, Vol. 26, N 4, 2013, p. 33–39.

[9] Malinowski M, Gopakumar K, Rodriguez J, Pérez MA. A survey on cascaded multilevel inverters. IEEE Trans. on Industrial Electronics, Vol. 57, N° 7, 2010, pp. 2197–2206.

[10] Le Besnerais J, Lanfranchi V, Hecquet M, Romary R, Brochet P. Optimal slot opening width for magnetic noise reduction in induction motors. IEEE Trans. on Energy Conversion 2009, Vol. 24, N 4, pp. 869–874.

[11] Brudny JF, Lecointe JP. Rotor design for reducing the switching magnetic noise of AC electrical machine variable-speed drives. IEEE Trans. on Industrial Electronics, Vol. 58, 2011, p. 5112–5120.

[12] Madarres M, Vahedi A, Ghazanchaei MR. Investigations on dynamic and steady state performance of axial flux hysteresis motors considering rotor configurations.

International Review of Electrical Engineering, Vol. 5, N 1, ISSN 1827-6660, 2010. p. 105.

[13] Da Silva ERC, Dos Santos EC, Jacobina CB. Pulsewidth modulation strategies. IEEE Ind. Electr. Mag, 5, 2, 2011, pp. 37–45.

[14] Kim DJ, Jung JW, Hong JP, Kim KJ, Park CJ. A study on the design process of noise reduction in induction motors. IEEE Trans. on Magnetics, Vol. 48, N 11, 2012, pp. 4638–4641.

[15] Habetler TG, Divan DM. Acoustic noise reduction in sinusoidal PWM drives using a randomly modulated carrier. IEEE Trans. Power Electr., Vol. 6, 1991, pp. 356–363.

[16] GA Covic, Boys JT. Noise quieting with random PWM AC drives. IEE Proc. Electr. Power Appl, Vol. 145, N1, January 1998, pp. 1–10.

[17] Le Besnerais J, Lanfranchi V, Hecquet M, Brochet P. Characterization and reduction of audible magnetic noise due to pwm supply in induction machines. IEEE Trans. Ind. Electr., Vol. 57, N 4, 2010, pp. 1288–1295.

[18] Ruiz-Gonzalez A, Meco-Gutierrez MJ, Perez-Hidalgo F, Vargas-Merino F, Heredia-Larrubia JR. Reducing acoustic noise radiated by inverter-fed induction motors controlled by a new PWM strategy. IEEE Trans. Ind. Electr., Vol. 57, N 1, 2010, pp. 228–236.

[19] Tischmacher H, Eichinger B. Sound optimisation of a converter-fed drive system using an acoustic camera in combination with modal analysis, COMPEL Vol. 29, N 4, 2010, p. 908.

[20] Szkudlapski T, Lecointe JP, Morganti F, Brudny JF. AC rotating machine radial vibrations: a principle to reduce the PWM switching effects. Proceedings of International Symposium on Electromagnetic Field, ISEF, Ohrid, Macedonian September 2013, PS5-271.

[21] Brudny JF, Szkudlapski T, Morganti F, Lecointe JP. Method for controlling the PWM switching: application to magnetic noise reduction. IEEE Trans. Ind. Electr., Vol. 62, N 1, 2015, pp. 122–131.

[22] Brudny JF, Morganti F, Lecointe JP, Parent G. On the use of carrier phase jumps to reduce some PWM switching effects. In Proceedings of IECON 2014—40th Annual Conf. on IEEE Industrial Electronics Society, Dallas, Texas, USA, October 2014, pp. 762–768.

[23] Alger PL. The nature of induction machines. 2nd ed, Gordon and Breach Publishers, New-York, London, Paris, 1970.

[24] Timar PL, Fazekas A, Kiss J, Miklos A, Yang SJ. Noise and vibration of electrical machines. Elsevier Amsterdam, Oxford, New York, Tokyo, 1989.

[25] Brudny JF. Modelling of induction machine slotting. Resonance phenomenon. European Physical Journal, Applied Physic, JP III, 1997, pp. 1009–1023.

[26] Lecointe JP, Romary R, Brudny JF, Czapla T. Five methods of stator natural frequency determination: case of induction and switched reluctance machines. Mechanical Systems and Signal Processing, MSSP, Academic Press, Cambridge, UK. 2004, pp. 1133–1159.

[27] Bowes SR. New sinusoidal pulse width-modulated invertor. Proceedings of the Institution of Electrical Engineers, Vol. 122, N 11, 1975, pp. 1279–1285.

[28] Holmes DG. A general analytical method for determining the theoretical harmonic components of carrier based PWM strategies. Proceedings of Thirty-Third IAS Annual Meeting, St Louis, MO, USA, IEEE Publisher, Vol. 2, October 1998, pp. 1207–1214.

[29] Sun J, Grotstollen H. Fast time-domain simulation by waveform relaxation methods, IEEE Trans. on Circuits and Systems I: Fundamental Theory and Applications, Vol. 44, N 8, 1997, pp. 660–666.

[30] Cassoret B, Corton R, Roger D, Brudny JF. Magnetic noise reduction of induction machines. IEEE Trans. on Power Electronics, Vol. 18, N 2, 2003, pp. 570–579.

Chapter 5

Tyre/Road Noise Annoyance Assessment Through Virtual Sounds

Elisabete Freitas, João Lamas, Carlos Silva,
Francisco Soares, Sandra Mouta and Jorge Santos

Additional information is available at the end of the chapter

http://dx.doi.org/10.5772/64359

Abstract

Road-traffic noise is the most significant source of environmental noise. Among the several different sources of noise emission from vehicles, tyre/road noise at speeds above 40 km/h is the most prevalent. Its negative impact on health is now better known and may be mitigated by optimising road surface characteristics. Experimental data linking the characteristics of the road surface to levels of annoyance regarding noise remain scarce. Moreover, assessing annoyance by experimental means using real sounds is complex and could impede study interactions with a wide set of variables. In this chapter, we describe, discuss and present the results of a straightforward method to assess tyre/road noise and related annoyance, based on the virtual sounds made by vehicles, with no interferences.

Keywords: traffic noise, annoyance, virtual noise, tyre/road noise

1. Introduction

Road-traffic noise is the most significant source of environmental noise. Among the several different sources of vehicle noise emission, tyre/road noise is predominant at a wide range of speeds, from low to high, but more so above 40 km/h. Its negative impact on health is now better known [1] and may be mitigated by optimising road pavement surface characteristics. Even though a lot of research and effort has been made in the last few years to develop and construct low-noise road pavement, annoyance remains a problem. Studies linking the road surface characteristics involved in tyre/road noise generation mechanisms with levels of annoyance are scarce.

Moreover, the field recordings in urban or rural environments are susceptible to contamination by other noise sources, thus limiting considerably the number and type of factors, which can be controlled and manipulated, or even the feasibility of the studies. Therefore, to avoid these drawbacks, a more straightforward procedure that combines noise acquisition close to the tyre (close proximity measurements) with auralisation of these acquisitions in a virtual scenario, tested and validated, would be useful. In addition, by manipulating the sounds measured in the source (the tyre), it will be possible to manipulate and analyse a wider set of traffic parameters and other road environment features.

This chapter aims to briefly introduce tyre/road noise generation mechanisms and analyse pavement surface characteristics and factors involved, and then to present and discuss the results of an innovative and straightforward method to assess tyre/road noise annoyance based on vehicles' virtual sounds generated from continuous close proximity tyre/road measurements.

2. Background

2.1. Road-traffic annoyance studies

The health impact of noise and of road-traffic noise, in particular, is commonly studied through annoyance assessments, either by surveys and questionnaires directed to large populations or by laboratory experiments with selected samples [2, 3].

Some of the most recent annoyance studies have been concerned with the health-related quality of life as, for example, studies on the annoyance caused by exposure to different noise sources, including road-traffic noise exposure [4], or studies on noise annoyance related to residential transportation and its impact on the physical activity of residents, change over time [5].

Taking into account the impact of the pavement surface on noise generation mechanisms, from low to high traffic speeds, it seems wise to comprehensively study the influence of road surfaces and their interaction with traffic characteristics on annoyance. From a practical perspective, the outcomes of these kinds of studies could be useful for highway engineers and decision makers. Nevertheless, previous studies have treated these annoyance factors separately. Some authors have been more concerned with road-traffic noise generation and addressed the impact of pavement type on annoyance [6, 7], while other authors addressed annoyance taking into account traffic parameters [8, 9], powered two wheelers and heavy vehicles and other factors such as industrial noise [10, 11].

In the scope of road-traffic noise generation, the most recent studies have analysed annoyance by taking into consideration the interactions of several types of pavements with relevant traffic parameters such as speed, type of vehicle and vehicle composition [12]. However, the main issue is that the experimental procedure to measure tyre/road noise is very complex and time consuming.

In an attempt to simplify the experimental procedure, the suitability of tyre/road noise close proximity measurements to assess annoyance through psychoacoustic parameters was examined [13, 14]. Furthermore, relations between subjective annoyance ratings and the traffic noise levels described by acoustic and psychoacoustic indicators (LA_{max}, LA_{eq} and Loudness) as a function of speed were established [14].

Despite the good results, the experiment conditions do not yet fully replicate those occurring near the road. However, the use of virtual sounds, built from close proximity records, to assess traffic annoyance is an almost unexplored solution, insufficiently documented in the literature.

2.2. Tyre/road noise and pavement surface characteristics

The noise emitted by a single vehicle is generated by three main sources: the engine, exhaust and transmission noise; the aerodynamic noise; and the noise generated by the interaction of the tyre with the road surface. The predominance of each one depends on vehicle operation conditions. It is recognised that tyre/road noise becomes predominant above 40 km/h [6]. Nonetheless, this limit has had a tendency to become lower due to the technological improvements in vehicles which have taken place over the last few years, such as the use of electrical engines.

Tyre characteristics and pavement surface characteristics play a crucial role in noise generation. The energy created by the interaction of the tyre with the pavement surface is radiated as sound. The tyre/road surface noise generation mechanisms are both vibrational, resulting from the impact and the adhesion of tyre treads on the surface, and aerodynamical, resulting from air vibrations around the tyre and in pavement and tyre surface cavities. There are also noise-related amplification or reduction mechanisms such as the horn effect, the acoustical and mechanical impedance of the surface and the tyre resonance [6].

The main characteristics of the tyre that affect noise are the tyre structure, the geometry, the tread pattern, the hardness of the rubber and the overall condition (wear and ageing) [6].

The noise generated by the contact of tyres on pavement surfaces is also influenced by a set of factors. There are three main road pavement surfaces: cement concrete, asphalt concrete and modular pavement. In general, they are made of mineral aggregates and a cementitious or asphaltic binder. Modular pavements may be made solely of natural stone. Despite their obvious differences, factors such as texture, porosity, aggregate gradation, type of bitumen and stiffness/damping [15] affect the interaction noise.

High macro-texture [16] and low mega-texture, negative texture, high porosity [17], small and regular mineral aggregates and low stiffness [17] or high damping [14, 18] are the main characteristics of a low noise surface. Asphaltic surfaces with bitumen modified with rubber [19] or with elastic aggregate substitute materials [20] also contribute to tyre/road noise reduction.

Therefore, dense asphalt concrete, stone mastic asphalt and surface dressings are the ones that generate more noise in contrast with double and single porous asphalt, thin layers and poroelastic surfaces [21–24].

Changes in pavement surface characteristics occur with time due to traffic wearing and weather exposition, generating more noise [19]. Weather-related factors such as the temperature of the surface and water on the surface also affect tyre/road noise. As the temperature increases, the tyre/road noise decreases [25, 26], while wet road surfaces always generate more noise than dry surfaces [27].

2.3. Tyre/road noise measurement methods

2.3.1. The statistical pass-by method

The statistical pass-by (SPB) measurement procedure, as described in ISO 11819-1: 1997, was designed to evaluate vehicle and traffic noise generated on different sections of road surface under specific traffic conditions.

In the SPB method, the maximum A-weighted sound pressure levels (SPLs) of a statistically significant number of individual vehicle pass-bys are measured at a specified roadside location together with the vehicle speeds (**Figure 1**). Each measured vehicle is classified into one of the three vehicle categories: 'cars', 'dual-axle heavy vehicles' and 'multi-axle heavy vehicles'. For each of the three speed ranges defined (low, medium and high speed categories) as well as for each of the three vehicle categories, a nominated reference speed is given. Each individual pass-by level, together with its vehicle speed, is recorded. This data is then analysed to determine the statistical pass-by index (SPBI).

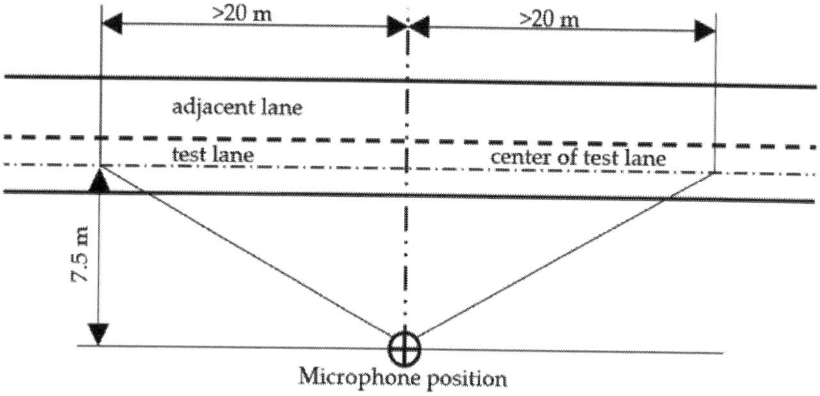

Figure 1. SPB testing geometry.

2.3.2. The controlled pass-by method

The controlled pass-by (CPB) method is a modified version of the statistical pass-by method that can be carried out using either a single vehicle or selected vehicles. In this method, the noise generated from a single car or light truck is measured when the vehicle approaches the test site at a specified speed in a specified gear.

2.3.3. *The statistical pass-by method using a backing board*

The statistical pass-by method using a backing board, ISO 11819-4:2013, specifies a modified version of the statistical pass-by method and uses a microphone mounted on a backing board instead of a microphone in normal, free-field conditions. It is suitable for measurements taken in an urban, built-up, environment or in the presence of safety barriers, noise barriers, embankments or road cuttings.

2.3.4. *The close proximity method*

In the 'close proximity (CPX) method', the average A-weighted SPLs emitted by specified tyres are measured over an arbitrary or a specified road distance, together with the vehicle testing speed, by at least two microphones located close to the tyres. For this purpose, a special test vehicle, which is either self-powered or towed behind another vehicle, is used (**Figure 2**). Reference tyres are mounted on the test vehicle.

(a)

(b)

Figure 2. Examples of CPX devices: at left (a) the interior of the TUG Tiresonic Mk4 CPX trailer [28]; at right (b) the University of Minho device [14].

According to ISO/DIS 11819-2, the tests are performed with the intention of determining a tyre/road sound pressure level, L_{CPX}, at reference speeds.

For each reference tyre and each individual test run with that tyre, the average sound pressure levels over short measuring distances, together with the corresponding vehicle speeds, are recorded. The sound pressure level of each segment is normalised to a reference speed by a simple correction procedure.

The CPX level, $L_{CPXt,V}$, is the resulting average sound pressure level for the two mandatory microphones at the reference speed (V) for the reference tyre (t). If more than one reference tyre is used, the Close Proximity Sound Index is calculated.

2.4. Perceptual noise indicators

Currently, the impact of road-traffic noise and also tyre/road noise is represented by the A-weighted equivalent mean average sound level (LA_{eq}) or the A-weighted maximum sound Level (LA_{max}). These acoustic indicators do not account for sound perceptual base items used by the human auditory system, such as loudness, roughness, sharpness and others [29]. Therefore, a comprehensive tyre/road noise annoyance study should include not only the objective noise indicators but also the perceptual indicators (subjective).

There are many models used in psychoacoustics for predicting the subjective sensation of loudness, sharpness and roughness.

Loudness is the attribute of auditory sensation in terms of which sounds may be ordered on a scale extending from soft to loud [29]. Loudness depends primarily upon the sound pressure of the stimulus, but also upon its frequency, waveform and duration. The 'loudness level' of a sound is defined as the sound pressure level of a 1 kHz tone in a plane wave and frontal incident that is as loud as the sound; its unit is 'phon' [29].

Loudness models can be divided into ones which use Bark auditory filters (or critical bands) and ones which use Erb auditory filters, and can also be divided into steady-state and dynamic models [30]. Steady-state models account for spectral effects on loudness, while dynamic models also account for the effect of auditory temporal integration on loudness, therefore they are better suited for time-varying signals.

Examples of steady-state models are Zwicker's model [29] and Moore, Glasberg and Baer's model (MGB) [31]. Glasberg and Moore (GM) [32], using Erbs, and Chalupper and Fastl (FC) [33], using Barks, are examples of dynamic loudness models.

Sharpness is a measure of the high frequency content of a sound (over 1100 Hz). The greater the proportion of high frequencies, the 'sharper' the sound [29]. A sound of sharpness 1 acum is defined as a narrow band-noise one critical band wide at a centre frequency of 1 kHz having a level of 60 dB [29].

There are several models to calculate sharpness, for example, Aures [34] or Zwicker and Fastl [29]. Zwicker and Fastl's model is a weighted centroid of specific loudness, while Aures's model is more sensitive to the positive influence of loudness on sharpness [30]. High frequencies generated by traffic are determined by aerodynamical noise generation mechanisms that make this indicator suitable to quantify their impact on annoyance.

Roughness is a complex effect that quantifies the subjective perception of rapid fluctuations (15–300 Hz) in the sound received by auditory filters [29]. The unit of measure is the asper. One asper is defined as the roughness produced by a 1000 Hz tone of 60 dB which is 100% amplitude modulated at 70 Hz [29].

Models of roughness for simple stimuli are given by Zwicker and Fastl [29]. For arbitrary stimuli, the roughness model optimised by Daniel and Weber [35] is applicable.

3. Method to assess traffic noise annoyance

3.1. Description of the method

The method to assess traffic noise annoyance follows four main steps: (I) tyre/road noise measurement, including road surfaces selection; (II) running an auralisation process to the acquired sounds using head-related transfer functions (HRTFs); (III) determining objective and subjective acoustic indicators; and (IV) running annoyance rating experiments **(Figure 3)**.

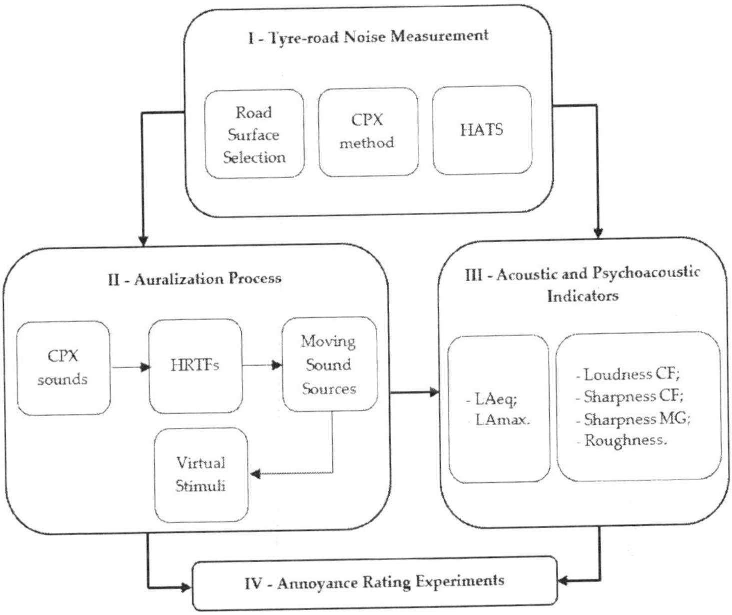

Figure 3. Organogram of the traffic noise annoyance method.

3.2. Noise measurement

Depending on the goal of the measurement, tyre/road noise measurement methods have advantages and disadvantages. SPB measurements include not only all vehicle noise sources and the effect of all propagation mechanisms, but also include other acoustic information near the measurement. For the evaluation of roadside noise through the SPB method, unreliable results due to the influence of surrounding conditions, even using a modified procedure, have been reported [36]. Records with head and torso simulator are in reality the SPB measurements, having precisely the same problems **(Figure 4)**. For annoyance or vehicle detection tasks, noise measurements contaminated with undesired sounds are an impediment to adequately accomplishing the task.

Figure 4. Measuring systems for SPB tests: microphone and HATS.

With the adoption of the CPX method, this important problem is eliminated. The influence of noise sources other than tyre/road is eliminated at most driving speeds. Also, this measurement method has been shown to be suitable for annoyance studies [37]. Moreover, since the source of tyre/road noise is in close proximity to the tyre/road interface, a substantial part of the propagation effect by acoustically absorptive surfaces is included in the microphone signal. In this way, the acoustic contribution of the pavement can be isolated and virtualised. Other effects, besides the pavement, can be added if desired.

3.3. Virtual noise construction

To create virtual noise, which is equivalent to the moving sound source obtained with the CPB measurements, an auralisation process was used. This technique enabled the virtual stimuli to be constructed using tyre/road CPX measurements as input sounds, resulting in the sound of a vehicle passing by the subject in a frontal-parallel plane.

All the stimuli were built with 5 seconds of length each. The vehicle passes the subject at the middle of the stimulus length. Like the CPB measurement, the position of the subject is located at 7.5 m away from the centre of the lane at 1.2 m height (see **Figure 5**). The auralisation process was made using Head Related Transfer Functions, and a geometrical attenuation was calculated considering a previous study [38].

http://dx.doi.org/10.5772/64359

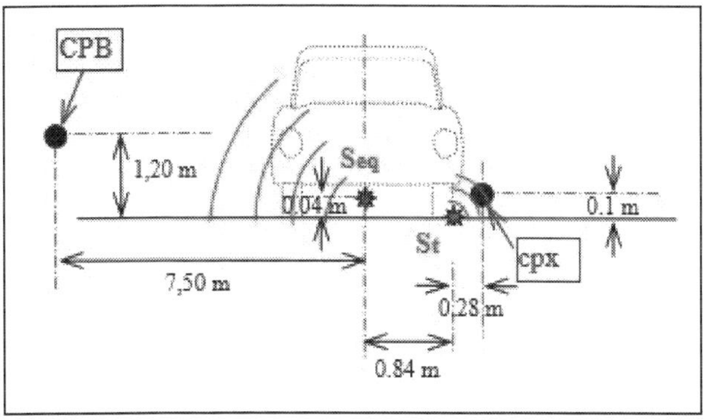

Figure 5. Noise sources location [38].

The original input sound, the CPX measurement sound, was partitioned into small samples; and for each sample, the correct attenuation level and the correspondent HRTF filter according to the azimuth angle between the subject and the vehicle position were applied. To apply the filter, head-related impulse responses (HRIRs) were used, convolved with the small samples in the time domain.

Depending on the vehicle speed during each trial, the vehicle's sound source position and azimuth were calculated for each consecutive small sample. For example, if the vehicle was travelling at 50 km/h, the initial position was at 69.45 m from the passing point, 69.85 m from the subject and the azimuth angle variation in this trial would be between −83.86 and +83.86 degrees. A Doppler effect was also applied since the stimuli consist of sound sources moving along a well-defined trajectory.

To calculate the geometrical sound attenuation, a sound pressure level difference between the CPX and the CPB measurement was defined:

$$\Delta L = L_{\mathrm{CPX}} - L_{\mathrm{CPB}} \tag{1}$$

Figure 5 represents a simplification of this model in which an equivalent point source is defined as S_{eq}, and this point was used in the virtual noise construction as the point moving along the previously defined trajectory.

The following equation represents the complete geometrical model:

$$\Delta L = Att_{\mathrm{CPX}} - Att_{\mathrm{ct}} - 20\log_{10}\left(\frac{r_{\mathrm{CPX}}}{r_{\mathrm{ct}}}\right) - 10\log_{10}\left(1+\left(\frac{r_{\mathrm{ct}}}{r_{\mathrm{ft}}}\right)^2 10^{\frac{Att_{\mathrm{ft}}-Att_{\mathrm{ct}}}{10}}\right) - 10\log_{10} \tag{2}$$

where Att_{CPX} is the sound pressure attenuation between the CPB measurement position and the S_{eq} position relative to the free field; Att_{ct} is the sound attenuation relative to free field between the closer tyres and the CPB measurement position; Att_{ft} is the sound attenuation relative to free field between the farther tyres and the CPB measurement position; r_{CPX} is the distance between the CPX measurement position and the tyre source position; r_{ct} is the distance between the closer tyres and the CPB measurement; and r_{ft} is the distance between the farther tyres and the CPB measurement.

Some noise contributions were not taken into consideration, such as the exhaust, engine and transmission noise as well as the sound reflections in the car body. The road surface attenuation was not simulated. Consequently, the stimuli reproduce a pure reflective road surface, and taking this into account, the Att_{ct}, Att_{ft} and Att_{CPX} have a value of 6 dB for all frequencies. The final equation of attenuation is defined as:

$$\Delta L = 10\log_{10}\left(\frac{r_{ct}^{2} r_{ft}^{2}}{2 r_{CPX}^{2}\left(r_{ct}^{2}+r_{ft}^{2}\right)}\right) \tag{3}$$

Since this equation represents the geometrical difference between L_{CPX} and L_{CPB}, a distance compensation was also used between the listener and the sound source for all the consecutive small samples in each sound.

3.4. Experimental method to measure traffic noise annoyance

Road traffic is an important source of noise annoyance and contributes to significant public health problems [39]. In terms of road safety, annoyance and detection should be studied in an integrated way, in order to understand and predict the effects of environmental noise on human behaviour and performance. Furthermore, these variables should be analysed and should be used as evidence when taking decisions in the planning and implementation of road infrastructures or vehicle features. Recent progress in this area has contributed to a reduction in road-traffic noise through the development of more efficient pavements and quieter vehicles, such as hybrid and electric cars. However, traffic noise plays a key role in road safety as it enhances the detection of other road users or vehicles, and consequently provides relevant information for road users and promotes the awareness of hazardous situations.

Previous works have demonstrated that pavement type and vehicle noise significantly affect the levels of environmental noise and associated subjective annoyance measurement [40] and detection levels [41]. Vehicle type affects both detection levels and annoyance ratings with a positive correlation between sound intensity, detection rates and annoyance. Regarding pavement type, detectability and annoyance rating revealed it to be more sensitive to this variable, in particular in lower speed traffic environments, such as those where hazardous situations with vulnerable road users might occur. In this context, a trade-off between detectability and annoyance levels should be addressed, accounting for the vehicle detection levels and overall traffic annoyance ratings.

As previously described, noise exposure can be described with physical measures that can be defined in different ways and be easily measured. Noise annoyance, in turn, is a subjective parameter considered to reflect the internal exposure to noise by associating the physical stimulus with the sensorial, perceptual and physiological consequences that it produces in the human organism.

From a methodological perspective, noise annoyance may be assessed either in field or in laboratory studies. Field studies are usually related to long-term annoyance and allow the study of factors encountered in real-life situations (e.g. noise sensitivity and noise source characteristics [42]).

Experimental methods in controlled environments allow the systematic manipulation of the traffic noise and related vehicle detection variables as a function of conditions such as pavement, vehicle type, background noise or user characteristics. The main goal of these kinds of methods is to quantify perceptual and/or behavioural performance as a function of the manipulation of physical or psychophysical (in this case, psychoacoustic) variables. Therefore, psychophysics can contribute substantially to the study of noise annoyance because it reduces the subjectivity of comfort or annoyance reports and allows the analysis of the effects of the variables on their own and the interactions between them [43, 44]. Each of the psychoacoustic indicators isolated might be unable to predict the annoyance on their own, but psychoacoustic metrics can correlate with non-sensory parameters and together can result in robust predictors or estimates [45]. Furthermore, simulated environment experiments can increase the ecological significance of the task and allow researchers to simulate real-life situations when studying the influence of acoustical and non-acoustical indicators on annoyance.

Based on the research goals and experimental design, requirements and criteria are defined (stimuli, space and material, procedure) to carry out the psychoacoustic experimental protocol in controlled environments. Prior to the experimental activities with human participants, an ethical approval from the local ethics committee will be required. Researchers should submit the selection criteria of the participants, experimental protocol and data management plan. For inclusion in the sample, participants are required to undergo an audiometric test to confirm that they have normal hearing. A learning phase or a familiarisation with the stimuli might be necessary for specific perceptual and detection tasks. Sometimes, the participant can freely hear the different stimuli and in other situations can, for example, compare each stimulus with the other sounds to set a relative scale.

The annoyance assessment of each participant is performed in a quiet room. Usually, sound stimuli are randomly presented together with channel reversed stimuli to avoid inter-aural biases following the adaptive or the constant stimulus method [46]. Participants are requested to assess the annoyance by responding to a Likert scale, between 5 and 11 points (from 'less annoying' to 'very annoying'). Some specifications and recommendations for this kind of instrument can be consulted in the ISO standard 15666:2003.

4. General outcomes

To support the previously discussed methodology, the main general outcomes of the research, which has already been carried out and of that which is also ongoing in this area, are examined. Two experiments in particular are addressed. One experiment was designed to verify if close proximity tyre/road noise measurements were suitable to rate traffic annoyance and the other one is a full application of the method described, aiming at verifying if virtual sounds generated from close proximity tyre/road noise measurements (virtual stimuli) can be used to rate annoyance correctly.

Some pavement surfaces were common to both studies, such as granite cubes (GC), concrete blocks (CB) and asphalt concrete (AC), and the testing speeds were 30 and 50 km/h. These conditions were used to illustrate hereafter the main results of the studies. Tyre/road noise measurement details can be found in [12] and [14].

The stimuli in both experiments were composed of the combination of pavement surface type with testing speed, and presented to voluntary listeners.

The annoyance assessment procedure of each participant was the same in both experiments. Each participant listened to the experimental noise trials and was requested to assess the annoyance of each noise trial with a 10-graded interval scale from 1 (less annoying) to 10 (very annoying).

The first approach to the problem consisted of checking if CPX annoyance ratings would have a similar trend to the CPB real annoyance ratings and then check if real annoyance ratings would have a similar trend to the virtual annoyance ratings. These trends were confirmed.

The mean annoyance rates, as expected, have the highest levels for the CPX stimulus, as they were recorded at the source, and lower levels for the virtual CPB stimuli, as they are deprived of engine, mechanical and background noises. This is clear for tyre/road noise coming from low noise surfaces. However, real ratings can be predicted by a simple model as a function of virtual stimuli annoyance rates and account for both the effect of speed and the type of pavement. In addition, using this methodology, it was possible to detect significant effects of type of pavement and speed on annoyance rates.

Psychoacoustic metrics can correlate with non-sensory parameters and together they can result in robust predictors or estimates. Good significant correlations of loudness with speed and pavement type have been found in previous studies. The final step to validate the method to assess annoyance using virtual sounds consisted of correlating the psychoacoustic metrics with the annoyance rates, aiming to use them as predictors. For the CPX annoyance prediction, there were no clear advantages for using loudness to predict annoyance. The fit of the annoyance rate against loudness was good for both acoustic and psychoacoustic indicators. However, for CPB annoyance rates, the quality of the prediction greatly improved for virtual stimuli and was better for loudness.

5. Discussion and conclusion

The methodology developed to assess tyre/road noise has been demonstrated to be efficient and will have a wide impact in the field, mostly in methodological and practical terms.

In general, the methodology significantly simplifies field procedures, allowing a reduction in experimental costs and a better control of variables, and it also brings great benefits with an increase in the accuracy of annoyance ratings.

On the one hand, the method examined allows researchers to fully achieve the objective of strictly assessing tyre/road noise annoyance by ignoring vehicle mechanical noises and generating virtual stimuli with tyre/road noise only. On the other hand, improvements in the simplified sound propagation algorithm can be made, for example, by including effects such as absorption and reflection. In addition, by adding engine noise, from electrical or traditional vehicles, and other noises, the masking effect of traffic annoyance could be better studied and understood.

There are several major advantages of using virtual instead of real stimuli; since besides the better control of experimental factors, there are also improvements in the quality of the recordings, and time consumption and costs are substantially reduced.

The most important practical impact of the method is the possibility to accurately establish requirements to control traffic noise based on indicators that better describe the annoyance triggered by the most significant source, the traffic.

Author details

Elisabete Freitas[1,2*], João Lamas[3], Carlos Silva[3], Francisco Soares[1,2], Sandra Mouta[3] and Jorge Santos[3,4,5]

*Address all correspondence to: efreitas@civil.uminho.pt

1 Centre of Territory Environment and Construction, University of Minho, Guimarães, Portugal

2 Department of Civil Engineering, School of Engineering, University of Minho, Guimarães, Portugal

3 Centre for Computer Graphics, Guimarães, Portugal

4 Centre Algoritmi, University of Minho, Guimarães, Portugal

5 Department of Basic Psychology, School of Psychology, University of Minho, Guimarães, Portugal

References

[1] WHO. Burden of disease from environmental noise. World Health Organization, Regional Office for Europe; 2011.

[2] Guski R., Felscher-Suhr U., Schuemer R. The concept of noise annoyance: how international experts see it. Journal of Sound and Vibration. 1999; 223(4):513–527.

[3] Ouis D. Annoyance from road traffic noise: a review. Journal of Environmental Psychology. 2001; 21:101–120.

[4] Héritier H., Vienneau D., Frei P., Eze I., Brink M., Probst-Hensch N., et al. The association between road traffic noise exposure, annoyance and health-related quality of life (HRQOL). International Journal of Environmental Research and Public Health. 2014; 11:12652–12667.

[5] Foraster M., Eze I., Vienneau D., Brink M., Cajochen C., Héritier H., et al. Annoyance to transportation noise and risk of physical inactivity. In: EuroNoise; 31 May-3 June; Maastricht; 2015.

[6] Sandberg U., Ejsmont J.A. Tyre/road noise reference book. INFORMEX; 2002. 640 p.

[7] Golebiewski R., Makarewicz R., Nowak M., Preis A. Traffic noise reduction due to the porous road surface. Applied Acoustics. 2003; 64:481–494.

[8] Griefahn B., Gjestland T., Preis A. Annoyance of residents living in urban areas. Report to partners Deliverable A.D6: month 36 SILENCE Integrated Project – Contract N. 516288: European Commission; 2008.

[9] Paunović K., Jakovljević B., Belojević G. Predictors of noise annoyance in noisy and quiet urban streets. Science of the Total Environment. 2009; 407:3707–3711.

[10] Gille L., Marquis-Favre C. Does the order of different successive vehicle pass-bys have an influence on the annoyance due to an urban road traffic noise? In: EuroNoise; 31 May-3 June; Maastricht; 2015.

[11] Morel J., Marquis-Favre C., Gille L.A. Noise annoyance assessment of various urban road vehicle pass-by noises in isolation and combined with industrial noise: a laboratory study. Applied Acoustics. 2016; 101:47–57. DOI: http://dx.doi.org/10.1016/j.apacoust.2015.07.017

[12] Freitas E., Oliveira J., Machado P. Innovative low noise road pavement materials studied in Portugal. The Journal of the Acoustical Society of America. 2012; 131(4):3225–3225. DOI: http://dx.doi.org/10.1121/1.4708026

[13] Licitra G., Cerchiai M., Teti L., Alfinito L. Road pavement description by psychoacoustical parameters from CPX data. In: 38th International Congress and Exposition on Noise Control Engineering. 2009, INTER-NOISE 2009; Canada; 2009. Pp. 1539–1545.

[14] Freitas E.F., Tinoco J., Soares F., Costa J., Cortez P., Pereira P. Modelling tyre-road noise with data mining techniques. Archives of Acoustics. 2015; 40(4):547–560. DOI: 10.1515/aoa-2015-0054

[15] SILVIA. Guidance manual for the implementation of low-noise road surface. FEHRL report 2006/02. Forum of European National Highway Research Laboratories, Brussels, Belgium; 2006.

[16] Bueno M., Luong J., Terán F., Viñuela U., Paje S.E. Macrotexture influence on vibrational mechanisms of the tyre-road noise of an asphalt rubber pavement. International Journal of Pavement Engineering. 2013; 15(7):606–613. DOI: 10.1080/10298436.2013.790547

[17] Liao G., Sakhaeifar M.S., Heitzman M., West R., Waller B., Wang S., et al. The effects of pavement surface characteristics on tire/pavement noise. Applied Acoustics. 2014; 76:14–23. DOI: http://dx.doi.org/10.1016/j.apacoust.2013.07.012

[18] Biligiri K.P., Way G.B. Noise-damping characteristics of different pavement surface wearing courses. Road Materials and Pavement Design. 2014; 15(4):925–941. DOI: 10.1080/14680629.2014.902768

[19] Freitas E.F. The effect of time on the contribution of asphalt rubber mixtures to noise abatement. Noise Control Engineering Journal. 2012; 60(1):1–8. DOI: 10.3397/1.3676311

[20] Bueno M., Luong J., Terán F., Viñuela U., Vázquez V.F., Paje S.E. Noise reduction properties of an experimental bituminous slurry with crumb rubber incorporated by the dry Process. Coatings. 2014; 4(3):602–613.

[21] Andersen B., Bendtsen H., Larsen L. Acoustic performance of low-noise road pavements. Silvia Project Report. Denmark; 2006.

[22] Descornet G., Goubert L. Noise classification of road pavements, Task 1: technical background information 1. Draft Report 05. Directorate-General Environment, European Commission; 2006.

[23] Bartolomaeus W. The potential of different road surface designs in road traffic noise reduction – European experience on pavement influence on noise (Experiences in Germany). Jornadas Técnicas; Madrid, Spain; 2006.

[24] Miljković M., Radenberg M. Thin noise-reducing asphalt pavements for urban areas in Germany. International Journal of Pavement Engineering. 2012; 13(6):569–578. DOI: 10.1080/10298436.2011.569028

[25] Bueno M., Luong J., Viñuela U., Terán F., Paje S.E. Pavement temperature influence on close proximity tire/road noise. Applied Acoustics. 2011; 72(11):829–835. DOI: http://dx.doi.org/10.1016/j.apacoust.2011.05.005

[26] Anfonsso-Lédée F., Pochaud Y. Temperature effect on tyre-road noise. Applied Acoustics. 2007; 72(11):829–835. DOI: http://dx.doi.org/10.1016/j.apacoust.2011.05.005

[27] Freitas E., Pereira P., de Picado-Santos L., Santos A. Traffic noise changes due to water on porous and dense asphalt surfaces. Road Materials and Pavement Design. 2009; 10(3):587–607. DOI: 10.3166/rmpd.10.587-607

[28] ROSANNE. Rolling resistence, Skid resistence, and Noise emition measurement standards for road surfaces [Internet]. Available from: http://rosanne-project.eu/wp/wp-content/uploads/The-TUG-Tiresonic-Mk4-CPX-trailer-interior.jpg [Accessed: 08-04-2016].

[29] Zwicker E., Fastl H. Psychoacoustics: Facts and Models. New York: Springer; 1999.

[30] Cabrera D., Ferguson S., Schubert E. 'Psysound3': software for acoustical and psycho-acoustical analysis of sound recordings. In: Proceedings of the 13th International Conference on Auditory Display; June 26–29; Montréal, Canada; 2007.

[31] Moore B.C.J., Glasberg B.R., Baer T. A model for the prediction of thresholds, loudness, and partial loudness. Journal of the Audio Engineering Society. 1997; 45:224–240.

[32] Glasberg B.R., Moore B.C.J. A model of loudness applicable to time-varying sounds. Journal of the Audio Engineering Society. 2002; 50:331–342.

[33] Chalupper J., Fastl H. Dynamic loudness model (DLM) for normal and hearing-impaired listeners. Acta Acustica United with Acustica. 2002; 88:378–386.

[34] Aures, W. Berechnungsverfahren für den sensorischen Wohlklang beliebiger Schall-signale. Acustica. 1985; 59:130–141.

[35] Daniel P., Weber R. Psychoacoustical roughness: implementation of an optimized model. Acustica. 1997; 83:113–123.

[36] Licitra G., Cerchiai M., Teti L., Ascari E., Bianco F., Chetoni M. Performance assessment of low-noise road surfaces in the Leopoldo project: comparison and validation of different measurement methods. Coatings. 2015; 5(1):3–25.

[37] Freitas E., Cunha C., Lamas J., Mouta S., Santos J.A. A psychoacoustic based approach to pavement classification. In: EuroNoise; 31 May–3 June; Maastricht; 2015.

[38] Anfosso-Lédée F. Modeling the local propagation effects of tire-road noise: propagation filter between CPX and CPB measurements. In: Inter.noise; 22–25 August; Prague; 2004.

[39] Ndrepepa A., Twardella D. Relationship between noise annoyance from road traffic noise and cardiovascular diseases: a meta-analysis. Noise Health. 2011; 13:251–259.

[40] Freitas E., Mendonça C., Santos J.A., Murteira C., Ferreira J.P. Traffic noise abatement: how different pavements, vehicle speeds and traffic densities affect annoyance levels. Transportation Research Part D. 2012; 17:321–326.

[41] Mendonça C., Freitas E., Ferreira J.P., Raimundo I.D., Santos J.A. Noise abatement and traffic safety: the trade-off of quieter engines and pavements on vehicle detection. Accident Analysis and Prevention. 2013; 51:11–17.

[42] Klaeboe R. Noise and health: annoyance and interference. Encyclopaedia of Environmental Health. 2011:152–163.

[43] Dittrich K., Oberfeld D. A comparison of the temporal weighting of annoyance and loudness. Journal of Acoustical Society of America. 2009; 126(6):3168–3178.

[44] Kryter K. Acoustical, sensory, and psychological research data and procedures for their use in predicting effects of environmental noises. Journal of Acoustical Society of America. 2007; 122:2601–2614.

[45] Stansfeld S., Kamp V.I., Hatfield J., Ellermeier W., Griefahn B., Lopez-Barrio I., et al. An examination of parametric properties of four noise sensitivity measures: research proposal. In: Inter.noise; Hawaii; 2006.

[46] Goldstein E.B. Sensation and Perception. 9th ed. Belmont, CA: Wadsworth Cengage Learning; 2013.

Section 2

Noise Mitigation and Control

Chapter 6

Open Acoustic Barriers: A New Attenuation Mechanism

Constanza Rubio, Sergio Castiñeira-Ibáñez,
Juan Vicente Sánchez-Pérez, Pilar Candelas,
Francisco Belmar and Antonio Uris

Additional information is available at the end of the chapter

http://dx.doi.org/10.5772/64360

Abstract

One of the main environmental problems of the industrialised countries is the noise, which can be defined as an unwanted or unpleasant outdoor sound generated by transport, industry and human activities in general. When it is not possible to reduce the emission of noise acting on the source, the reduction of noise levels in its transmission phase using acoustic screens (AS) seems appropriate; such screens are in common use to reduce noise levels and have been extensively studied since the middle of the 20th century. Over the last decades, various acoustic screen designs have been investigated to increase the screening effect. The research carried out focuses on both the reduction of diffraction at the top edge of the screen by varying the shape at the top or adding absorptive materials to the noise screen, but all these screens are basically formed by a continuum rigid material with a superficial density high enough, to reduce transmission of noise through the screen, in accordance with the mass law. At the end of the nineties, another type of screen based on arrangements of isolated scatterers embedded in air, emerged. Among other interesting properties, these screens provide new mechanisms to control the noise based on the Bragg law. First, a Sonic Crystal Acoustic Screen (SCAS) was presented, where the scatterers are arranged following crystalline patterns. After that, a new prototype of AS based on sonic crystals appears, which increases the attenuation capabilities using arrangements based on fractal geometries. The screens designed in this way have been referred to as Fractal-based Sonic Crystal Acoustic Screens (FSCAS) in this chapter. In both the cases, the mechanism that prevents the transmission of noise, and therefore increases the noise attenuation, is the destructive Bragg interference due to a multiple scattering process. Finally, a new concept of AS based on a periodic arrangement of scatterers, with a slit dimension between them that is smaller than the wavelength is introduced. This latest screen is called Subwavelength Slit Acoustic Screen (SSAS) which presents a Wood anomaly and Fabry-Perot resonances, being the destructive interferences among the scattered waves, responsible for the attenuation capabilities of these screens. This new kind of AS (SCAS, FSCAS and SSAS) presents interesting properties and can be considered as a real alternative to the classical AS, which are formed by a continuum rigid material. The aim of this chapter is to present these open AS, and it is organised as follows. In Section 1, an introduction about classic acoustic screens is presented. Numerical models and experimental set-up for the screens are introduced in Section 2. Then, in Section 3 the transmission properties of Sonic Crystals are explained, and the research advances in this field related to the design of a screen based on the new mechanism of noise control are highlighted. The definition and development of the Fractal-based Sonic Crystal Acoustic Screen are shown subsequently. The Subwavelength Slit Acoustic Screen is developed in Section 4. Finally, in Section 5 the main results and conclusions of the work are presented.

Keywords: acoustic barrier, sonic crystal, subwavelength slit, acoustic attenuation, insertion loss

1. Introduction

Noise pollution is an important environmental problem of the twenty-first century society that affects millions of people around the world. In urban areas where noise pollution is due to different noise sources, such as traffic noise, industrial noise and many other noise sources, the citizens are exposed to high noise levels, which have been linked to various health problems such as fatigue, sleep disturbance, cardiovascular disorders and/or some reduction on performance at work or school [1, 2]. In general, noise travels from the source to the receiver in a straight line, so that we can act on both the source (reducing noise generated by motors and machinery, for example) on the receiver (installing more insulation enclosures such as double windows) or on the noise transmission phase (installing acoustic screens) to control the noise that reaches the receiver [3]. When it is not possible to reduce the emission of the noise acting on the source or on the receiver—technically or economically—the most effective and common method to reduce the noise levels is the use of acoustic screens (AS). A classical AS is defined as any relatively opaque solid to sound that blocks the line of sight from the sound source to the receiver. In order to satisfy this, screens have to be made with continuum rigid materials (without openings) since the procedure which reduces sound transmission is the mass law, as discussed above, the superficial density must be at least 20 kg/m^2. The acoustic mechanism involved in this kind of screen is simple: The interposed object (the screen) reduces the noise levels that reach the receiver through diverse mechanisms such as absorption, reflection or diffraction (see **Figure 1**). This latest mechanism means that the acoustic wave, rather than following the direct path, it is forced to spread over or around the screen. This mechanism can be considered as one of the main factors that decreases the effectiveness of the screens [4]. The zone where sound waves do not arrive through the straight path is called the shadow zone. Here the sound pressure level is mainly due to diffraction.

Many researchers [5–9] have studied the performance of AS with different top patterns. Apart from this research line, uninteresting advances in this research file have been obtained, leading to a stagnation in the technological development in the field of AS.

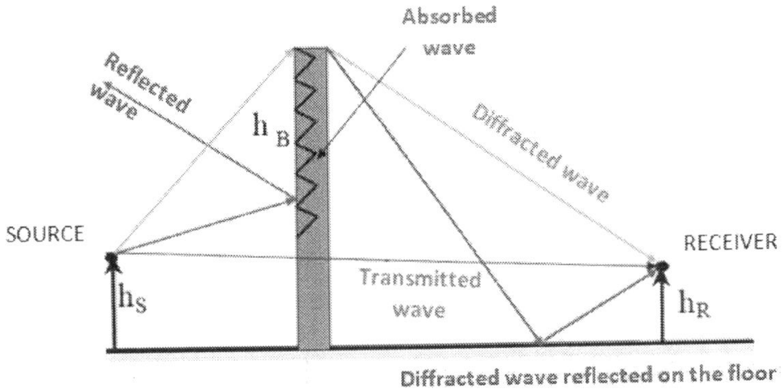

Figure 1. Noise path and acoustic behaviour of sound when an acoustic screen is placed between source and receiver.

At the end of the nineties, a new research line focused on alternative screen systems has received considerable attention. This chapter reports the modelling, development and testing of this new kind of screens that includes another physical mechanism to control the transmission of acoustic waves. Non-acoustic factors such as cost, practical and engineering issues, aesthetic or environmental impacts must be taken. The need for an AS may be mostly driven by acoustic concerns; however, non-acoustic factors are as important as the acoustic ones in the final design and construction.

Finally, it must be pointed out that the index commonly used to express the acoustic attenuation obtained by installing AS is the insertion loss (IL). This index is defined as the difference between the sound pressure levels recorded at the same point with and without the screen. Throughout this chapter, IL will be used to standardise and compare the acoustic behaviour of the different devices presented.

2. Numerical model and experimental results

The development of theoretical models that explain the interaction of acoustic waves with different objects is one of the fundamental pillars in the development of acoustics, allowing us to understand both the underlying physics in new systems and devices, as well as studying their potential technological applications. In order to predict the performance of an AS before fabrication, numerical calculations have been made. Due to the characteristics of the devices analysed, the Finite Element Method (FEM) seems a good theoretical tool to design this kind of noise screen. Thus, a commercial software Comsol Multiphysics 3.5a, Acoustics Pressure Module (Time Harmonic Analysis) is used for modelling the different screens proposed. A free and fine mesh parameter, with the regular refinement method was selected.

The wave equation, partial differential equation (PDE), is given by the expression [10]:

$$\nabla\left(\frac{1}{\rho_0}\nabla p\right)+\frac{\omega^2}{c^2\rho_0}p=0$$

In order to analyse the behaviour of the screen a plane wave impinging on it is considered. Throughout this chapter, and for all the new screens defined, we only consider the normal incidence of the wave on the structure. This plane wave can be expressed as follows:

$$p_{in}=e^{j\omega x}$$

The noise attenuation spectrum, at a point behind the sample, is obtained for the different cases. To do that, the difference between the direct and interfered sound pressure evaluated by means the Insertion Loss (IL) parameter at that point, has been calculated as follows:

$$IL=20\cdot log_{10}\left|\frac{P_{direct}}{P_{Interfered}}\right|$$

Concerning boundary conditions used in this model, it is had [10]:

- *Sound-Hard Boundary (Wall):* This is a boundary condition of the Neumann type, implying that the partial derivative of the pressure at the surface is zero. This means that the amplitude of the pressure at this surface will be maximum or minimum.
- *Radiation Boundary Condition:* It allows the wave travelling out the modelling domain, with minimal or no reflection, that is, the scattered field consists only of outgoing waves. This condition is also called the Sommerfeld condition.
- *Perfectly Matched Layers (PML):* in situations where it is not possible to describe in a simple way, the outgoing radiation with a known wavenumber and its propagation direction, PMLs offer a powerful alternative: while a PML has the same function, it is not a boundary condition but an additional domain that absorbs the incident radiation without producing reflections. PML provides good performance for a wide range of angles of incidence being insensitive to the shape of the wave fronts. PML therefore is able to emulate the limits avoiding reflections.

The height of the scatterers for all the screens defined is considered infinite, so the diffraction at the top edge was not considered and the structure can be considered by means of a two dimensional (2D) model.

By a way of example, in **Figure 2**, these conditions and the possibility to obtain pressure maps and the attenuation spectrum are shown. The structure, which is placed in the numerical domain, will be any of the screens shown afterwards such as SCAS, FSCAS or SSAS, because all of them can be considered formed by rigid scatterers.

http://dx.doi.org/10.5772/64360

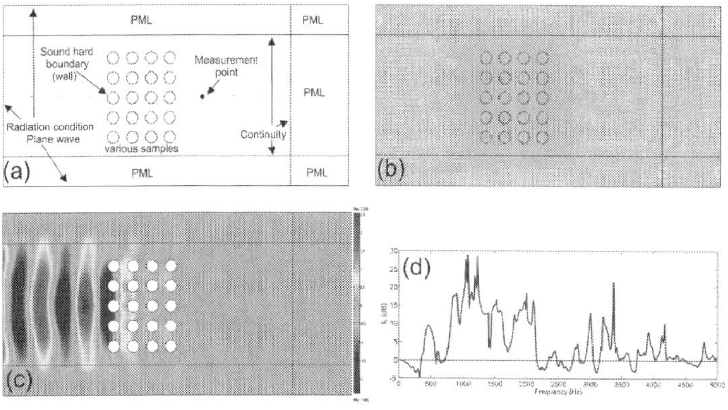

Figure 2. (a) Main conditions of the numerical solution domain. (b) Numerical domain mesh of 2.5 × 10^5 elements. (c) Numerical simulation of the scattering problem at 1000 Hz. The real value of the total pressure, Re(P), is plotted. (d) Attenuation spectrum at a selected point.

To validate the numerical results obtained by means of our model, a set of experiments have been performed in an echo-free chamber of size 8 × 6 × 3 m^3, which simulate free field conditions [11]. A prepolarized free-field 1/2″ microphone Type 4189 B&K located at 1 m from the sample and a directional sound source GENELEC 8040A emitting continuous white noise, located 1 m behind the sample in order to consider the wave impinging on the sample as a plane wave, have been used throughout the experiments. The position of the microphone is controlled and varied by using a Cartesian robot, which we have called the three-dimensional Robotized e-Acoustic Measurement System (3DReAMS), which is capable of sweeping the microphone through a 3D grid of measuring points located at any trajectory inside the chamber. When the robotized system is turned off, the microphone acquires the temporal signal. This signal is saved on the computer and then the frequency response of the measured sample is obtained using the Fast Fourier Transform (FFT). For both the data acquisition and the motion of the robot, the National Instruments cards PCI-4474 and PCI-7334 have been used together with the Sound and Vibration Toolkit and the Order Analysis Toolkit for LabVIEW. The sample was hanging from a frame in such a way that there was no influence of ground effect. Experimental results shown in this chapter have been obtained, basically, by means of this experimental set-up. The special features will be commented in sections 3 and 4.

3. Sonic Crystal Acoustic Screen (SCAS)

3.1. Main characteristics on sonic crystals

The first option presented as an alternative to classical acoustic screens is the use of sonic crystals (SCs). A SC can be defined as a heterogeneous material consisting of a periodic array of acoustic scatterers embedded in a medium with different physical properties, where one of the materials is a fluid [12]. Depending on the number of directions where the periodicity of the array exists, one, two or three-dimensional SCs can be defined (see **Figure 3**). The mechanism that forbids the sound transmission through the system is different from those that prevail in classical AS, in this case the underlying physical mechanism is the destructive Bragg interference due to a Multiple Scattering (MS) process [13], related to the periodicity of the system. These destructive interferences lead to attenuation peaks. These peaks widen with an increase of both the number and the diameter of the cylinders, thus these periodic systems exhibit ranges of frequencies related to the periodicity of the structure where there is no wave propagation. These ranges of frequency are called Band Gaps (BGs). Other physical mechanisms such as absorption or reflection can be added to the scatterers increasing the attenuation capabilities.

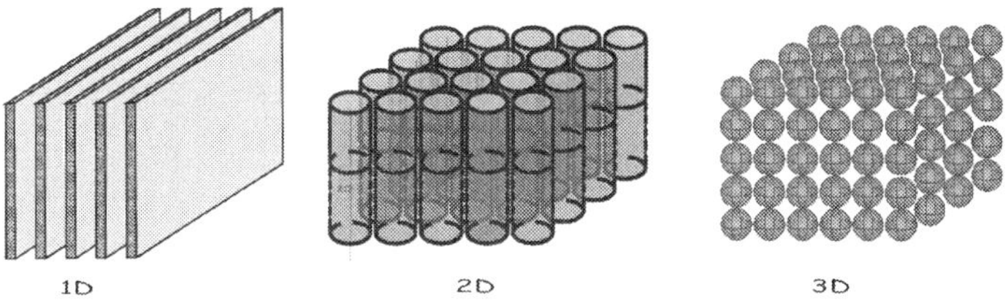

Figure 3. One, two and three-dimensional Sonic Crystal.

The use of SCs as AS implies that the scatterers are formed by rigid materials and the fluid medium is air. Furthermore, although the shape of the scatterers is related to the attenuation properties of these systems [14], the most used in the two-dimensional case is the cylindrical shape, mainly due to the ease of calculation associated with its geometry because of its symmetry properties. Therefore, we will consider two-dimensional sonic crystals (2DSC) formed by cylindrical scatterers with variable height and radius, and the screens formed in this way are called Sonic Crystal Acoustic Screens (SCAS).

Regardless of the shape, the scattering of the incident wave in each and every one of the scatterers of the SC leads to the existence of the BGs. The existence of the BGs is one of the most interesting properties of SCs and it is the responsible for their use as AS.

Both the position and the size of these BG depend, in the frequency domain, on several factors [15] such as: (i) the spatial distribution of the scatterers; (ii) the separation between the scatterers; (iii) the densities and the contrast of propagation velocities between the air and the scatterers and (iv) The amount of mass of the scatterers per unit area of the crystalline array. Next, the influence of these factors on the existence of the band gaps is briefly explained.

i. Spatial distribution of scatterers. This factor indicates how the scatterers are arranged in the crystal. Thus, for instance, if the chosen symmetry is triangular, scatterers are located at the vertices of the triangle; if the symmetry is square, at the corners of squares, and so on. The most suitable crystalline symmetries for 2DSC screens are square and triangular lattices.

ii. Separation between scatterers. This factor is given by the lattice constant and indicates the size of the triangles, squares, etc., which form the array under consideration. It is closely related to the previous factor because if the spatial distribution of scatterers defines how they are distributed; the lattice constant determines how much these scatterers are separated from one another. Furthermore, it is crucial in such a periodic system because it defines the relationship between the geometrical properties of the lattice and the position of the attenuation band in such a way that when the separation decreases, the position of the attenuation band in the frequency domain increases and vice versa. Therefore, by increasing the lattice constant very low frequencies could be attenuated.

iii. The densities and the contrast of propagation velocities between the air and the scatterers determine the size of the BGs and their position in the frequency domain. In our case, the material that forms the scatterers must be acoustically rigid; with its propagation velocity and its density values being much higher than those of air. This fact allows the use of almost any material in the design of the scatterers.

iv. The amount of mass of the scatterer per unit area of the crystalline array. This relationship is given by the filling fraction, which represents the relation between the area occupied by scatterers and the area of the lattice. In general, the higher the filling fraction the broader the BG (see **Figure 4**).

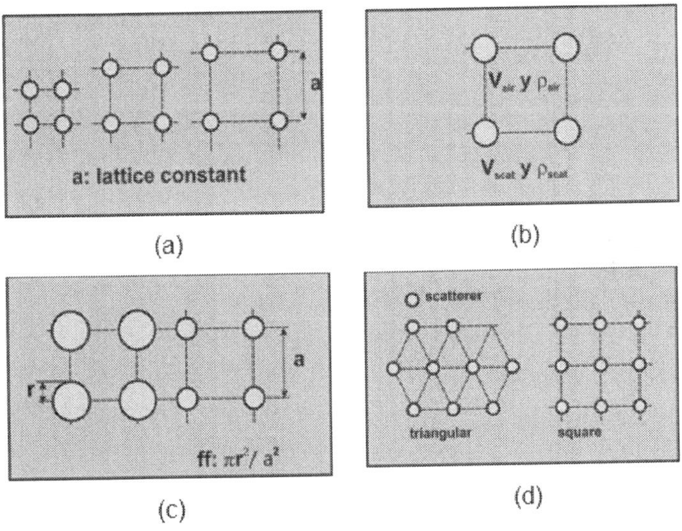

Figure 4. Main parameters that determine both the position and the size of the BGs. (a) Separation between scatterers (lattice constant). (b) The densities and the contrast of propagation velocities between the air and the scatterers. (c) The amount of mass of the scatterers per unit area of the crystalline array (ff). (d) Spatial distribution of scatterers.

Moreover, the position of the BG in the frequency domain depends on the direction of incidence of the acoustic wave on the SC.

Another parameter commonly used to quantify the noise attenuation in a range of frequencies is the attenuation area (AA), defined as the area enclosed between the positive attenuation spectra and the 0 dB threshold in the frequency range selected [16].

Symmetry	Bragg's frequency	Filling fraction
Square	$f_{Bragg} = \frac{c}{2\,a}$	$ff_{square} = \frac{\pi\, r^2}{a^2}$
Triangular	$f_{Bragg} = \frac{c}{\sqrt{3}\,a}$	$ff_{triangular} = \frac{2\,\pi\, r^2}{\sqrt{3}\,a^2}$

Table 1. Mathematical expressions for determining the Bragg's frequency and filling fraction for square and triangular symmetries. Cylindrical scatterers are considered as having *r* for its radii and *a* is the lattice constant.

Where the scatterers are cylindrical and sound impinges perpendicularly onto the screen, the Bragg frequencies and the filling fraction for square and triangular symmetry are those summarised in **Table 1**.

In **Figure 5a**, the attenuation spectrum of a SC formed by 60 cylindrical scatterers with radii r = 0.09 m arranged in a square lattice with lattice constant a = 0.22 m is represented. With these characteristics, the Bragg frequency of the attenuation peak is given as follows:

$$f_{Bragg} = \frac{c}{2a} = 772\,Hz$$

The frequencies where the wavelength is approximately twice the lattice constant, form part of the first BG. In this figure, the first and the second BGs can be observed. In **Figure 5b**, the attenuation spectrum of a SC formed by 60 cylindrical scatterers with radii $r = 0.02$ m arranged in a triangular lattice with lattice constant $a = 0.0635$ cm is represented. In this case, the Bragg frequency is given as follows:

$$f_{Bragg} = \frac{c}{\sqrt{3}a} = 2677\,Hz$$

In both the cases SCs are formed by 6 rows each of 10 cylinders each one, with the incident wave (IPW) being normal to the structure.

As it noted above, the position of the BG in the frequency domain depends on the direction of incidence of the acoustic wave on the SC. However, if wide BGs are achieved an overlap between bands for different orientations can exist. In the overlap area, the screen attenuates regardless of the direction of incidence. In summary, the distance between the scatterers and the angle of incidence of the wave on the sonic crystal determines the central frequency of the BG, and the section size of the scatterers determines the width of frequencies of the attenuation bands. Throughout this chapter the incidence is considered normal to the structure.

It is worth noting that with the Bragg interference being the only mechanism to control the sound attenuation it is difficult to ensure a good design and construction of an efficient acoustic screen based on sonic crystals. To avoid this problem and also improve the attenuation properties of the SCAS, two research lines have been followed. On one hand, to avoid the dependence with the angle of incidence of the acoustic wave on the structure and, on the other, to maximise the destructive interference. For the first one, additional physical properties have been introduced to the scatterers such as absorption and resonances [17]. For the second one, novel arrangements of scatterers which are different from the crystalline symmetries that allow the overlapping of acoustic bands have been sought. This second research line is further developed in this chapter.

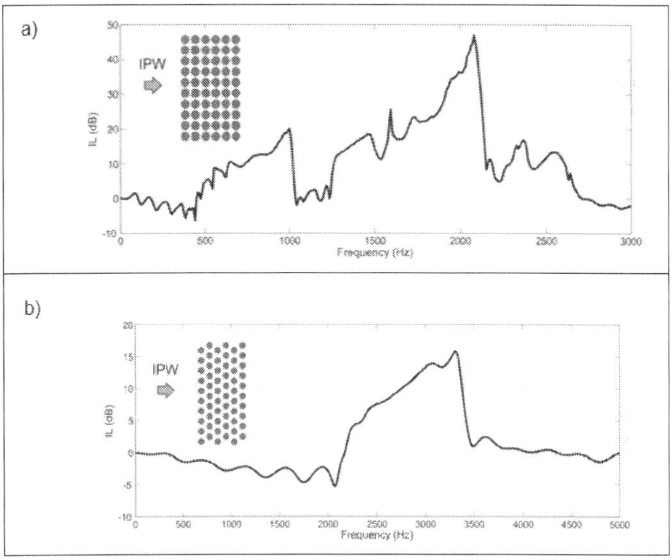

Figure 5. Attenuation spectra for normal incidence. (a) Square lattice, lattice constant a=22 cm and cylinder radii r=9 cm is represented. (b) Triangular lattice, lattice constant a=6.35 cm and cylinder radii r=2 cm.

3.2. Fractal-based Sonic Crystal Acoustic Screen

In nature there are objects whose structure has the property of scale invariance, that is, the whole seems geometrically equal to the parts. The objects that contain copies of themselves within them are usually called fractals. Fractal geometries have been applied in several sciences, such as medicine, biology [18] or economics [19]. In the case of SC, fractals have been traditionally used to design the shape of the scatterers [20] or to reallocate the total mass of the scatterers in a predetermined device [21].

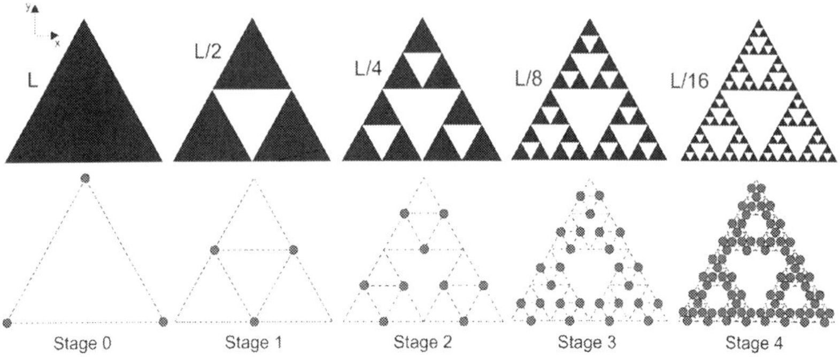

Figure 6. Quasi-fractal arrangement of scatterers: Five stages of cylinder arrays based on Sierpinski's triangle geometry.

This section provides a procedure based on the redistribution of the mass of the scatterers of SC following fractal geometries in order to increase the attenuation bands obtained by Bragg diffraction [21].

For this purpose, a well-known fractal called the Sierpinski triangle, whose formation can be seen at the top of **Figure 6**, has been used to design a two dimensional structure of isolated rigid cylinders with a high attenuation performance. As can be seen at the bottom of **Figure 6** the new arrangement of the scatterers is formed by placing them following this fractal geometry. Bearing in mind that fractal construction results from an infinite iterative process, that is, it results from the repetition of an identical theme at different size scales [22], the structure showed in **Figure 6** is called Quasi-Fractal Structure (QFS) due to the fact that only the first five stages or iterations are shown [21].

Moreover, QFS could be considered as classic triangular arrangements in which some cylinders have been removed, creating some vacancies in the array. However, the underlying symmetry follows a fractal pattern. Therefore, QFS can be considered as a sum of several triangular arrays with a different lattice constant (Remember that the lattice constant has been introduced in Section 2 and it is the distance between scatterers), such as L, L/2, L/4, L/8 and L/16 as shown in **Figure 6**, corresponding to each fractal stage. This procedure provides small and compact devices where the whole attenuation band obtained can be considered as the sum of the BG of each triangular array that forms the QFS.

In **Figure 7**, a cross section of the new arrangement is shown (in the plane OXY), considering cylinders of radius *r* parallel to the Z-axis. Although QFS could be considered as a classical triangular array with the same vacancies, one has to take into account that the underlying growth is fractal.

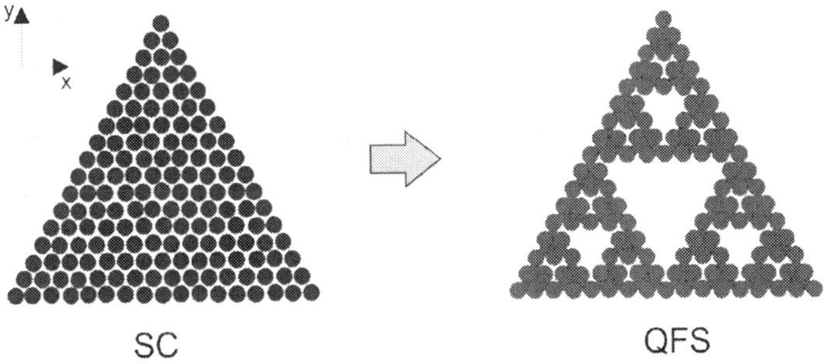

Figure 7. Starting SC and resulting QFS structures.

With this process, a compact device has been obtained. Then as a second step, in order to maximise the BG based on the quasi-fractal arrangement obtained, the radii of the scatterers have been varied at each stage. Bearing in mind that the efficiency of the ff is what is being searching, biggest radii were chosen for stage 1 and then, as the stage increases, the radius decreases. According to this second step, the most important features of this design technique will be explained right after.

For this maximisation process, the golden proportion was used to design the radii of the cylindrical scatterers. In **Figure 8**, the quasi-fractal proposed QFS_{Max}, which were constructed in accordance with the golden proportion among the radii of the cylinders of stages s (s = 0, 1, 2, 3, 4) is shown. The radius at each stage is listed in **Table 2**.

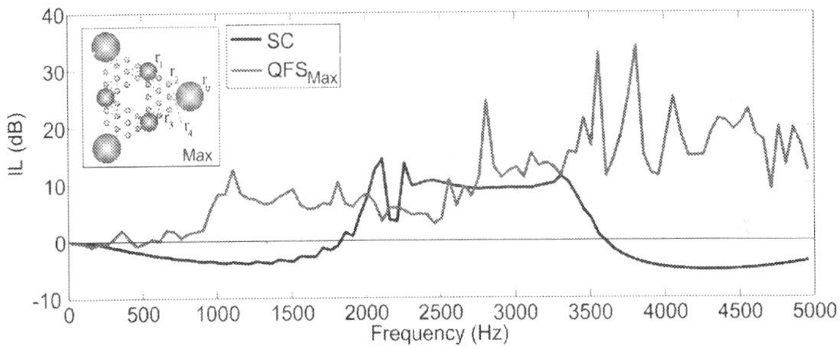

Figure 8. Sound attenuation spectra (IL) at normal incidence for both the starting SC and QFS_{Max}. In the inset, one can see a section of the quasi-fractal designed.

r_0	r_1	r_2	r_3	r_4
14.8 cm	9 cm	5.6 cm	3 cm	1.9 cm

Table 2. Theoretical radii of the of the cylinders of stages s (r_i) following the golden proportion.

It should be noted that it was necessary to remove some of the cylinders of the starting QFS shown in **Figure 8**, in order to place the larger cylinders (larger radius) of the, first iterations. It is worth noting that for other applications, the relationships between the radii of the cylinders could vary.

Comparing the values of the AA parameter for QFS_{Max} and a SC with similar size and external shape, made of rigid cylinders (radius, $r = 2$ cm) and arranged in a triangular lattice ($a = 6.35$ cm), interesting results can be observed: the value of the AA parameter for QFS_{Max} (AA_{Max} = 51048 Hz·dB) grows by over 340% compared to the starting SC (AA_{SC} = 14705 Hz·dB).

To build the QFS_{Max} it is necessary to adapt to the sizes of commercial scatterers. With this in mind, a QFS_{Exp} has been constructed with the available cylinders, with the radii in this case being listed in **Table 3**.

r_0	r_1	r_2	r_3	r_4
10	8	5.5	3	1.75

Table 3. Commercial radii of the of the cylinders of stages s (r_i) for QFS_{Exp}.

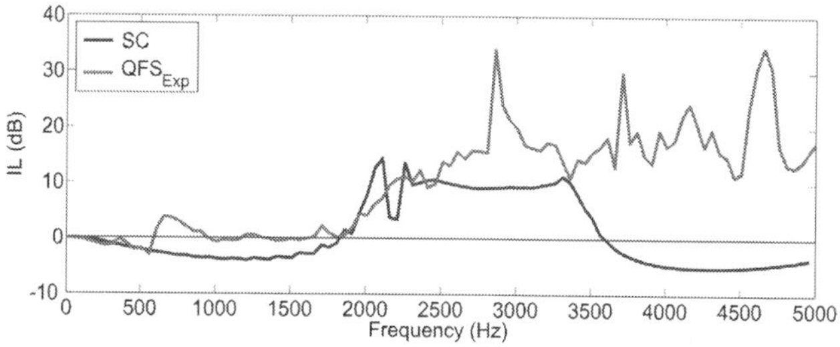

Figure 9. Spectrum sound attenuation (IL) at normal incidence to the starting SC and QFS_{Exp}.

Figure 9 shows the IL spectra for the case of commercial scatterers QFS_{Exp} and the starting SC, obtaining very interesting numerical results.

Thus, by comparing the values of the AA parameter for QFS_{Exp} (AA_{Exp} = 42391 Hz·dB) and the starting SC (AA_{SC}=14705 Hz·dB), by over than 280% of enhancement is obtained for the QFS_{Exp}. **Table 4** summarises the results:

Sample	No. of scatterers	ff (%)	AA (Hz·dB)
SC	136	36	14,705
QFS_{Max}	60	45.8	51,048
QFS_{Exp}	60	33.24	42,391

Table 4. Number of scatterers, filling fraction (ff) and attenuation area (AA) for SC, QFS_{Max} and QFS_{Exp}.

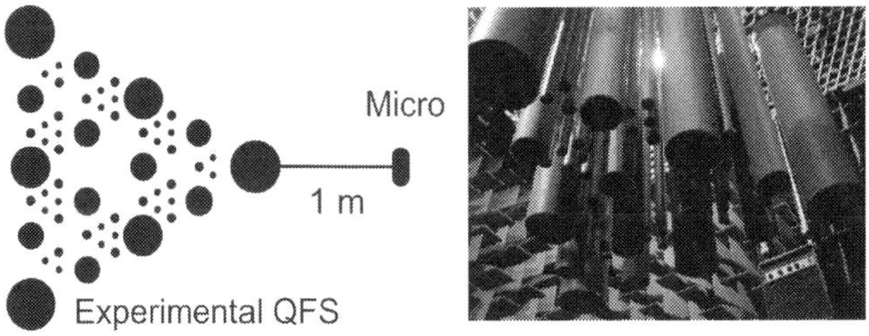

Figure 10. QFS experimental and QFS structure picture, taken from below the structure in an anechoic chamber.

With these results, the rule of the bigger ff the bigger BG size is broken. QFS produces wider BG band even though the ff decreases comparing the values with respect to the starting SC.

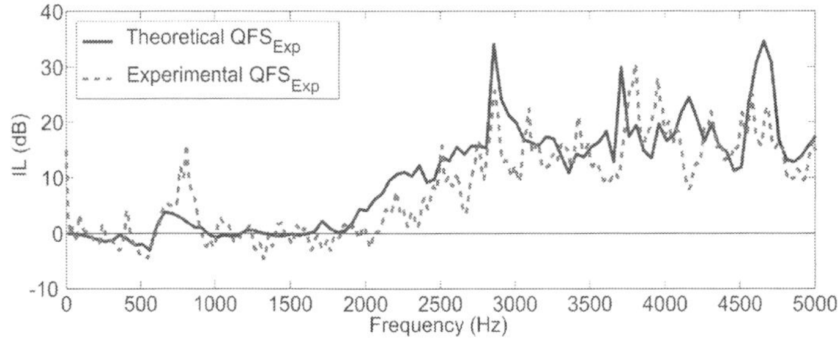

Figure 11. QFS_{Exp} theoretical and experimental IL.

To demonstrate these results, experimental measurements under controlled conditions have been performed in an anechoic chamber, as explained in Section 2.

In **Figure 10**, the QFS structure analysed and a photograph of the same within the anechoic chamber is shown.

In **Figure 11**, the experimental acoustic attenuation spectrum with QFS_{Exp} having been experimentally and numerically obtained is overlapped. The close agreement between both results can be appreciated, thus validating the design carried out.

By way of conclusion, the technique developed in two steps enables us to maximize the BG at normal incidence. At the first step the scatterers are arranged according to fractal geometry and secondly, the radii of the scatterers that form this fractal structure are adjusted depending on the considered stage in order to maximize the ff at each stage.

The screens developed from the QFS_{Exp} are the FSCAS. Nowadays, there are screens designed according to this technique that have even been homologated for market placement [23].

4. Subwavelength Slit Acoustic Screen

In 1998, Ebbesen et al. [24] observed extraordinary optical transmission through metallic films perforated with subwavelength aperture arrays. Since then, this new area of study has been the subject of research. Motivated by the wave nature of electromagnetic waves and acoustic waves, the findings for optic waves were transferred to acoustics and named "extraordinary acoustic transmission" through subwavelength apertures. In 2007, Lu et al. [25] showed both theoretically and experimentally the extraordinary acoustic transmission through a 1-D grating with subwavelength slits and Hou et al. [26] reported experimentally the extraordinary acoustic transmission through a subwavelength hole array. It is accepted that the extraordinary acoustic transmission through a subwavelength aperture array is motivated by the Fabry-Perot resonances inside the apertures. In 2008, Estrada et al. [27] showed, both theoretically and experimentally, that Wood anomalies [28] were the responsible for an extraordinary shielding in perforated plates with periodic subwavelength hole arrays. The position in the spectrum of the sound attenuation peaks corresponding to the Wood anomalies depends on the lattice constant and geometry of the array. Motivated by the success of the extraordinary shielding, a new acoustic screen based on subwavelength slits had been designed as an alternative to classical acoustic and SC screens. The sound attenuation capabilities of the new screen can be tuned as a function of the geometrical parameters, as in the case of SCAS.

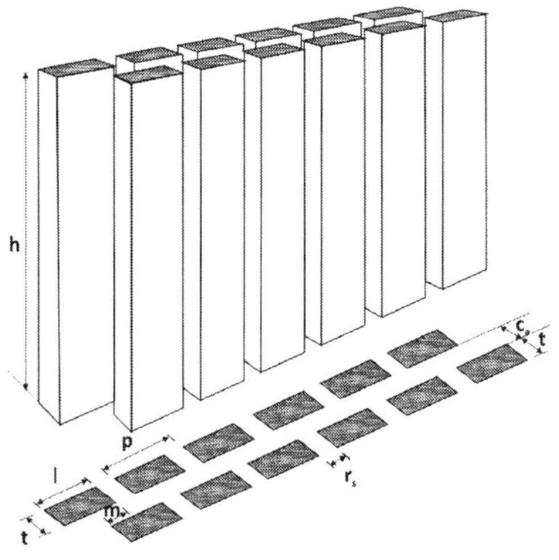

Figure 12. Schematic diagram of the subwavelength slit acoustic screen.

The basic structure considered has two rows of rigid pickets of length l and thickness t, distributed periodically with a lattice constant and with a slit width r_s. The rows are separated by an air gap c_a and the lateral misalignment between the rows is m as shown in **Figure 12**.

http://dx.doi.org/10.5772/64360

The effectiveness of the subwavelength-slit screen was evaluated in controlled conditions described in Section 2.

The pickets were made of wood with length $l = 0.3$ m, thickness $t = 0.1$ m and height $h = 1.8$ m, distributed periodically with a lattice constant $p = 0.35$ m, with a slit width $r_s = 0.05$ m. The rows were separated by an air gap $c_a = 0.1$ m. The picket dimensions and the periodicity were chosen to obtain the maximum attenuation corresponding to Wood anomalies at 1000 Hz.

The acoustic attenuation properties of this screen are represented by means of IL, covering a range frequency from 100 to 2500 Hz in 6 Hz steps.

In order to ensure that the position of the sound attenuation peak in the frequency domain corresponds to the Wood anomalies, several measurements for a single row of periodic array of pickets with the same characteristics as those used for the SSAS have been carried out. **Figure 13** shows the comparison between measured and calculated IL at normal incidence. Numerical computations were performed by means of FEM. The boundary conditions have been described in Section 2. The only difference with respect to **Figure 2a** is the structure that is located in the domain, in this case instead of cylinders, pickets must be placed. Pronounced insertion loss peaks can be observed, that is, the exact manifestation of the Wood anomaly [28]. The Wood anomaly for normal incidence in a periodic array of subwavelength slits is given by $\lambda = p$, where λ is the wavelength and p is the lattice period. The positions of the Wood anomalies in the frequency domain for a periodic array with p = 0.35 m were 971 and 1942 Hz. The Wood anomaly at the frequency of 1942 Hz is not clearly observed due to the interference with the Fabry-Perot resonance [29].

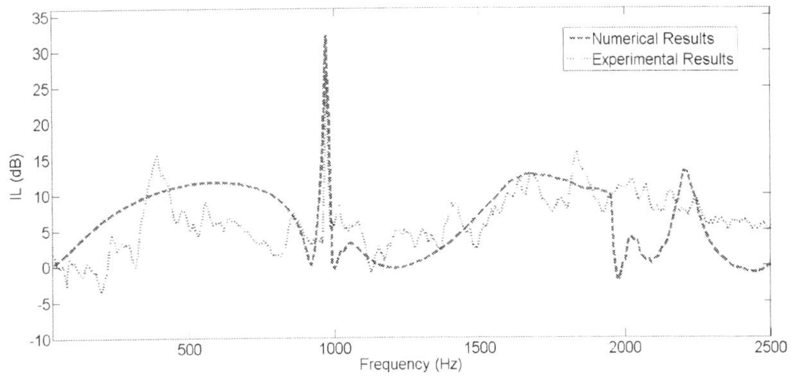

Figure 13. Calculated and measured IL at normal incidence for a single row of periodic array of pickets.

Figure 14 shows measured IL at normal incidence for the SSAS with the characteristics mentioned above. The rows were separated by an air gap $c_a = 0.1$ m. Three different values of lateral misalignment between the rows were considered $m = 0.3$, 0.2 and 0.125 m. The Wood anomaly at a frequency of 971 Hz is clearly seen. Sharp insertion loss peak at a frequency of around 1500 Hz is also observed. This frequency corresponds to destructive interference between the odd and even Fabry-Perot modes, resulting in an IL increase.

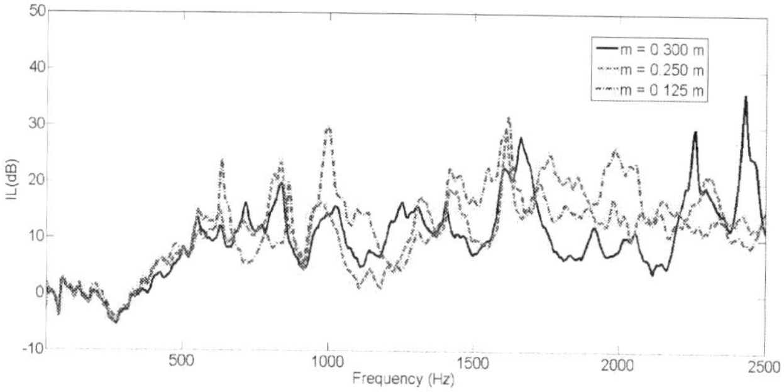

Figure 14. Measured insertion loss for an acoustic screen with two rows of pickets of length l = 0.3 m, thickness t = 0.1 m and height h = 1.8 m, distributed periodically with a period p = 0.35 m, with a slit width r_s = 0.05 m. The rows were separated by an air gap c_a = 0.1 m. Three different values of lateral misalignment between the rows were considered m = 0.300, 0.250 and 0.125 m.

By varying the lateral misalignment between the rows m, it is observed that, as the lateral misalignment is increased, insertion loss peaks appear at frequencies between 500 and 800 Hz which, in the case of lateral misalignment m = 0.125 m, did not appear. These peaks are associated with a destructive interference between the propagating and evanescent waves.

Thus, the values of the AA parameter for SSAS varying the lateral misalignment between the rows are shown in **Table 5**. With only two rows of pickets significant values of the AA parameters are achieved.

Misalignment (m)	AA (Hz·dB)
0.300	72,733
0.250	78,696
0.125	68,523

Table 5. Attenuation area (AA) for different values of lateral misalignment between the rows were considered.

5. Summary and conclusions

New types of screens based on arrangements of isolated scatterers embedded in air were presented in this chapter. First, a SCAS was presented, where the scatterers are arranged following crystalline patterns. After that, a new prototype based on sonic crystals but arranged following fractal geometries has been presented. The attenuation capabilities of these screens,

FSCAS, were enhanced by varying the filling fraction of each fractal stage separately. Finally, SSAS is introduced. Unlike the other ones, this latest screen is not based on crystalline geometry but consists of arrangements of isolated scatterers. The attenuation capabilities of these screens could be enhanced by introducing additional physical properties to the scatterers.

Author details

Constanza Rubio[1], Sergio Castiñeira-Ibáñez[2], Juan Vicente Sánchez-Pérez[1], Pilar Candelas[1], Francisco Belmar[1] and Antonio Uris[1*]

*Address all correspondence to: auris@fis.upv.es

1 Physical Technologies Center, Universitat Politecnica de Valencia (Polytechnic University of Valencia), Valencia, Spain

2 Electronic Engineering Department, Universitat de Valencia (University of València), Burjassot, Valencia, Spain

References

[1] Commission of the European Communities. Green Paper on "Future Noise Policy". Brussels: 1996. p. COM(96)540.

[2] Burden disease from environmental noise: quantification of healthy life years lost in Europe. World Health Organization Regional Office for Europe, Cpenhagen, Denmark. 2011

[3] Harris C. M., editor. Handbook of Acoustical Measurements and Noise Control. 3rd ed. New York: McGraw-Hill; 1991.

[4] Kotzen B., English C., editors. Environmental Noise Barriers: A Guide to Their Acoustic and Visual Design. 1st ed. London and New York: E & FN SPON; 1999.

[5] Hothersall D.C., Crombie D.H., Chandler-Wilde S.N. The performance of T-profile and associated noise barrier. Applied Acoustics. 1991;32:269–281.

[6] Okubo T., Fujiwara K. Efficiency of a noise barrier with an acoustically soft cylindrical edge for practical use. Journal of the Acoustical Society of America. 1999;105:3326–3334.

[7] Watts G.R. Acoustic performance of multiple-edge noise barrier profile at motorway sites. Applied Acoustics. 1996; 37:44–66.

[8] Shao W., Lee H.P., Lim S.P. Performance of noise barriers with random edge profiles. Applied Acoustics. 2001; 62:1157–1170.

[9] Byung-Joo J., Hyun-Sil K., Hyun-Ju K., Jae-Seung K. Sound diffraction by a partially inclined noise barrier. Applied Acoustics. 2001; 62:1107–1121.

[10] COMSOL Multiphysics User Guide: Version 3.5a, COMSOL AB, Stockholm, Sweden. 2008

[11] Castiñeira-Ibañez, S., Rubio, C., Sánchez-Pérez, J.V. Acoustic wave diffraction at the upper edge of a two-dimensional periodic array of finite rigid cylinders. A comprehensive design model of the periodicity-based devices. Europhysics Letters. 2013;101:64002.

[12] Martinez-Sala R., Sancho J., Sánchez-Pérez J.V., Gómez V., Llinares J., Meseguer F. Sound attenuation by sculpture. Nature. 1995; 378:241.

[13] Chen YY., Ze Z. Theoretical analysis of acoustic stop bands in two dimensional periodic scattering array. Physical Review E. 2001; 64:036616.

[14] Min R., Wu F., Zhong L., Zhong S., Liu Y.Y. Extreme acoustic band gaps obtained under high symmetry in 2D phononic crystals. Journal of Physics D: Applied Physics. 2006; 39:2272–2276.

[15] Kushwaha M. Stop-bands for periodic metallic rods:Sculptures that can filter the noise. Applied Physics Letters. 1997; 70:3218.

[16] Romero-Garcia V., Fuster E., García-Raffi L.M., Sánchez-Pérez E.A., Sopena M., Llinares J., Sánchez-Pérez J.V. Band gap creation using quasiordered structures based on sonic crystals. Applied Physics Letters. 2006; 88:174104.

[17] Romero-García V., Krynkin A., García-Raffi L.M., Umnova O., Sánchez-Pérez, J.V. Multi-resonant scatterers in sonic crystals: locally multi-resonant acoustic metamaterial. Journal of Sound and Vibration. 2013; 332:184–198.

[18] Iannaccone P.M., Khokha M.. Fractal Geometry in Biological Systems: An Analytical Approach. CRC Press; 1996.

[19] Willians B., Trading B.. Chaos, Applying Expert Techniques to Maximize Your Profits. Market place Books; 1995.

[20] Norris R.C., Hamel J.S., Nadeau P. Phononic band gap crystals with periodic fractal inclusions: theoretical study using numerical analysis. Journal of Applied Physics. 2008;103:104908.

[21] Castiñeira-Ibañez S., Romero-García V., Sánchez-Pérez J.V. García-Raffi L.M. Overlapping of acoustic bandgaps using fractal geometries. Europhysics Letters. 2010;92:24007.

[22] Mandelbrot B. The Fractal Geometry of the Nature. New York: W.H. Freeman & Co; 1983.

[23] Castiñeira-Ibáñez S., Rubio C., Romero-Garcia V., Sánchez -Pérez J.V., García-Raffi L.M. Design, manufacture and characterization of an acoustic barrier made of multi-

http://dx.doi.org/10.5772/64360

phenomena cylindrical scatterers arranged in a fractal-based geometry. Archives of Acoustics. 2012;37:455–462.

[24] Ebbesen T.W., Lezec H.J. Ghaemi H.F., Thio T., Wolf P.A. Extraordinary optical transmission through sub-wavelength hole arrays. Nature. 1998; 391:667–669.

[25] Lu M.H., Liu X.K., Feng L., Li J., Huang C.P., Chen Y.F., Zhu Y.Y., Zhu S.N., Ming N.B. Extraordinary acoustic transmission through a 1d grating with very narrow apertures. Physical Review Letters. 2007;99:174301.

[26] Hou B., Mei J., Ke M., Wen W., Liu Z., Shi J., Sheng P. Tuning Fabry-Perot resonances via diffraction evanescent waves. Physical Review B. 2007; 76:054303.

[27] Estrada H., Candelas P., Uris A., Belmar F., Garcia de Abajo F.J., Meseguer F. Extraordinary sound screening in perforated plates. Physical Review Letters. 2008;101:084302.

[28] Wood R.W. Anomalous diffraction gratings. Physical Review. 1935; 48:928–936.

[29] Rubio C., Candelas P., Belmar F., Gomez-Lozano V., Uris A. Subwavelength slit acoustic metamaterial barrier. Journal of Physics D: Applied Physics. 2015; 48:395501.

Chapter 7

Effects of Platform Screen Doors on Sound Fields in Underground Stations

Yoshiharu Soeta and Yong Hee Kim

Additional information is available at the end of the chapter

http://dx.doi.org/10.5772/64627

Abstract

This chapter investigates the acoustic effects of platform screen doors (PSDs) in underground stations using computer simulation and scale model testing. The dimensions of underground stations with island and side platforms were determined based on a field survey. Ray-tracing-based computer models and 1/25 scaled-down physical models of these underground stations were used to simulate their sound field characteristics. In the experiments, five types of PSDs were tested: mobile closed full-height (MCFH), mobile open full-height (MOFH), mobile half-height (MHH), fixed half-height (FHH) and fixed barrier (FB) doors. Four acoustic parameters, namely, speech intelligibility, sound pressure level, reverberation time and the inter-aural cross-correlation coefficient were used to understand the sound field characteristics from the sound source of public address announcements. It was found that speech intelligibility and the sound pressure level were increased by most types of PSDs apart from the MCFH. The MOFH showed the highest levels of speech intelligibility and spatial diffusivity. In addition, the noise reduction effects of PSDs for train noise were discussed. PSDs on side platforms showed higher noise reduction performances than PSDs on island platforms. The specific noise reduction levels for the MOFH type were 4.3 dB on island platforms and 5.0 dB on side platforms.

Keywords: platform screen door, underground station, computer simulation, scale model testing, speech intelligibility

1. Introduction

Noise in train stations can annoy the passengers, reduce the speech intelligibility on public address systems in the stations [1–3] and also pose the risk of causing noise-induced hearing loss in passengers, transit workers and operators [4–6]. The control of noise in train stations is thus important for the comfort, convenience and safety of passengers, transit workers and station staff. However, relatively few acoustical treatments are conducted in most train stations.

The architectural conditions of train stations are diverse, and their sound fields differ according to the station type. The effects of the location, i.e. whether the station is located aboveground or underground, and the platform style, which can be side (two platforms at the side and the rail tracks at the centre) or island (one platform at the centre and rail tracks on both sides), on the train noise in stations have been investigated recently [7]. The results show that the noise level in underground stations is 6.4 dB higher than that in aboveground stations, and that noise level in underground stations with island platforms, which are the most widely used types of stations in Japan for economic reasons, is higher than those with side platforms. This indicates that acoustic treatments are necessary for underground stations with island platforms.

Recently, platform screen doors (PSDs) have been widely applied in the platform areas in train stations for passenger safety in Asian and European metro systems [8, 9]. PSDs help to prevent falls from the platform onto the track area, reduce the risk of accidents being caused by service trains passing through stations at high speeds and improve the climate control, e.g. heating, ventilation and air conditioning, within the station. Railway companies in Japan are also encouraged to install PSDs for safety reasons. There are approximately 9500 train stations in Japan and approximately 4% of these stations have installed PSDs. More than 200 accidents leading to injury or death occurred in these stations in 2008. Over a 2-year period from 2008 to 2010, 61 train stations installed PSDs. The number of accidents that occurred in stations subsequently decreased by approximately 10%.

PSDs have a positive impact on safety in train stations, and they also have an effect on the train noise in stations because they are a type of wall that blocks the noise generated by trains. Some studies have reported that the acoustic environments in underground stations were improved by isolating the train noises [7, 10–14]. The strongest effect of the use of PSDs is the reduction of the train noise level. Reported noise reduction levels produced by the presence of PSDs in underground stations ranged from 9 to 18 dB in field measurements [10–14]. In the case of underground stations in the Seoul Metro, in South Korea, the equivalent sound pressure levels of the train noise from the arrival to departure process in stations with PSDs were approximately 71 dB(A), whereas those in stations without PSDs ranged from approximately 78 to 82 dB(A) [10]. The differences between the *in situ* measured sound pressure levels inside and outside PSDs were reported to range from 16 to 18 dB(A) [11]. In addition, reinforcement of early reflections for sounds from the public address (PA) systems was also found [9].

The changes in the sound field characteristics caused by the PSDs have rarely been studied. In particular, the contributions of PSDs to the improvement of speech intelligibility are yet not known. Also, the noise reduction effects of PSDs in a previous study [14] were derived from field measurements in different train stations because it was impossible to change the PSDs, after they have been installed. Different train stations have different acoustic characteristics, including volume, reverberation and background noise. Also, many different types of trains, which have different noise source characteristics, run through train stations that were measured, and thus it was impossible to eliminate the effects of the different noise sources in the field measurements. Therefore, the noise reduction performances of different PSDs can be predicted more accurately through a simulation approach using reliable acoustic models and different platform styles.

This study therefore investigated changes in the sound field characteristics including the noise reduction effects for different types of PSDs using computer simulations and acoustic scale modelling. It was hypothesized that the speech intelligibility and noise reduction performance levels would be improved by the presence of PSDs, depending on the type of platform shape.

2. Methods

2.1. Target underground stations

The architectural conditions of train stations are diverse, and their sound fields differ according to the station type. In terms of location, an aboveground station covered only by a roof is similar to a free sound field, while a completely covered underground station is a reverberant sound field [15]. To focus on the effects of reflections from the PSDs, the walls and the ceilings, only underground stations were simulated in this study.

In terms of platform style, stations can be principally divided into side (two platforms at the side and rail tracks at the centre) and island (one platform at the centre and rail tracks on the sides) platforms. In underground stations, the train runs down the centre of a station with side platforms, while it runs against the lateral wall of a station with an island platform. These different architectural elements can also change the characteristics of the sound fields.

Figure 1 shows the two simplified underground stations with island and side platforms that were selected for the study. Full details of the target stations, including their acoustic fitting to real stations, have been described in previous studies [7, 14, 16]. The platform length for both stations was 150 m, with a corridor height of 3 m. The width of a single track was 3.6 m. The distance between the track and the platform floors was 1.1 m. The maximum station height for both stations was 5.3 m at the position of the middle of the track. The maximum station width was 14.8 m for the island platform-type station, and 17.2 m for the side platform-type station. The cross-sectional area was 61 m^2 for the island platform-type station, and 68.2 m^2 for the side platform-type station. Both stations were simulated without either passengers or background noise.

2.2. Platform screen doors

Figure 2 shows the five types of PSDs that were used in this study: three mobile (MCFH: closed full-height, MOFH: open full-height, MHH: half-height) and two fixed (FHH: half-height, FB: barrier) types of PSD. The PSD dimensions were determined based on practical designs. The doors and the lower walls of the PSDs were made from tempered glass. The upper walls of the PSDs were made from sheet metal.

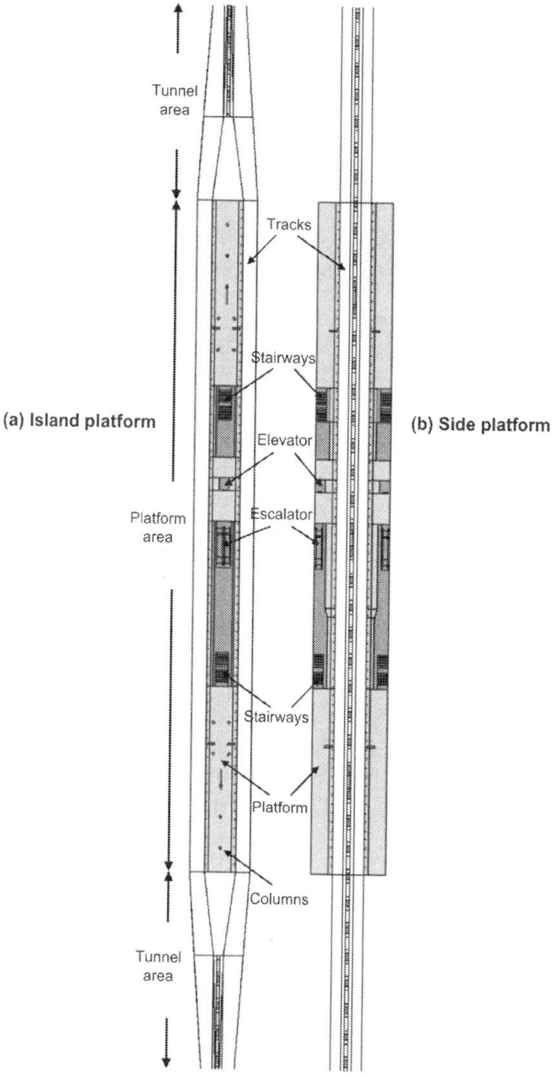

Figure 1. Floor plan of the target underground stations with (a) island and (b) side platforms.

2.3. Simulation models using the ray-tracing method

Ray-tracing software (Odeon 11.23) was used to derive the acoustic parameters and the binaural impulse responses. As shown in **Figure 3**, a total of 12 cases were simulated for the various PSD configurations, including the no-PSDs condition (NSD). For the simulation parameters, the transition order was 1 with 120,000 rays. The environmental conditions were 20°C and 50% relative humidity (RH).

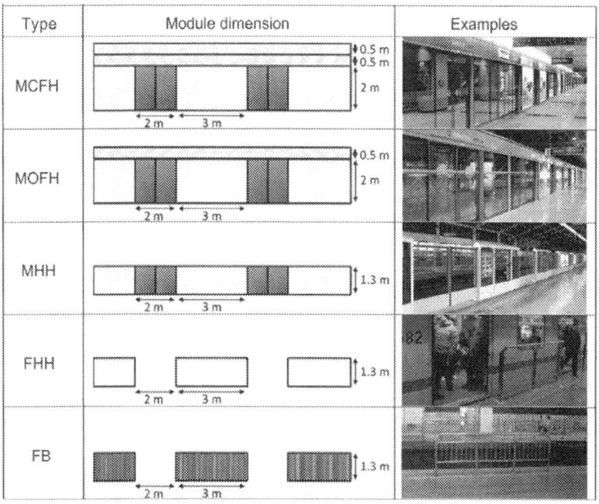

Figure 2. Modules of the five types of PSDs with their dimensions.

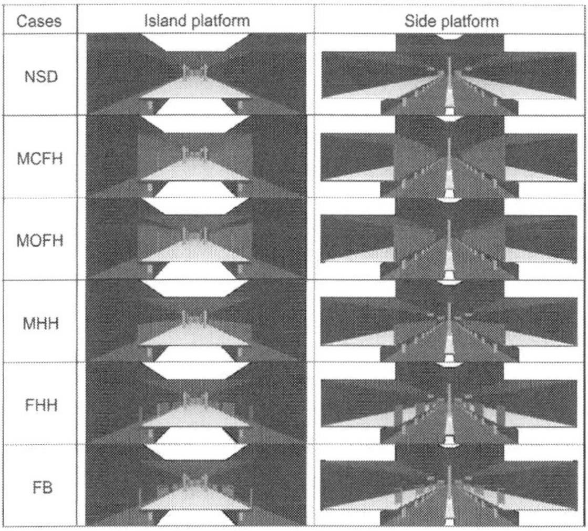

Figure 3. Simulation configurations according to the types of PSDs.

2.4. Physical models on 1/25 scale

A 1/25 scale model station with an island platform was built to validate the computer simulation results. The main body of the scale model was made from 9-mm-thick medium-density fibreboard with a varnish coating. The PSDs were made from 1-mm-thick Foamex plastic board (a type of polyvinyl chloride plastic). **Figure 4** shows the model testing configurations for the different types of PSDs.

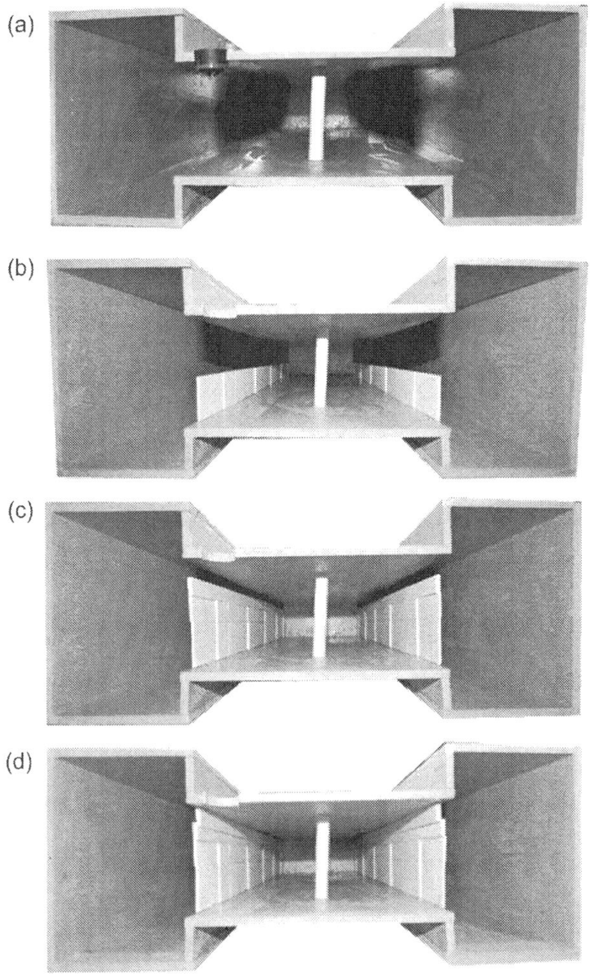

Figure 4. Section of the scale model stations with PSDs (a) NSD, (b) MCFH, (c) MOFH and (d) MHH.

2.5. Source and receiver position

In this study, two measurement configurations were used for the investigations. **Figure 5** shows the source and receiver positions for Configuration 1 in both the simulation and scale models for evaluation of the sound field characteristics to determine the speech intelligibility of PA sounds. A sound source at a height of 2.8 m (0.112 m in the scale model) was located 22.5 m (0.9 m in the scale model) away from the rear wall of the platform in the longitudinal direction. In total, 14 receivers were placed at a height of 1.6 m (0.064 m in scale model) on either side of the sound source in the longitudinal direction. The distance between the receivers was 2.5 m (0.1 m in the scale model).

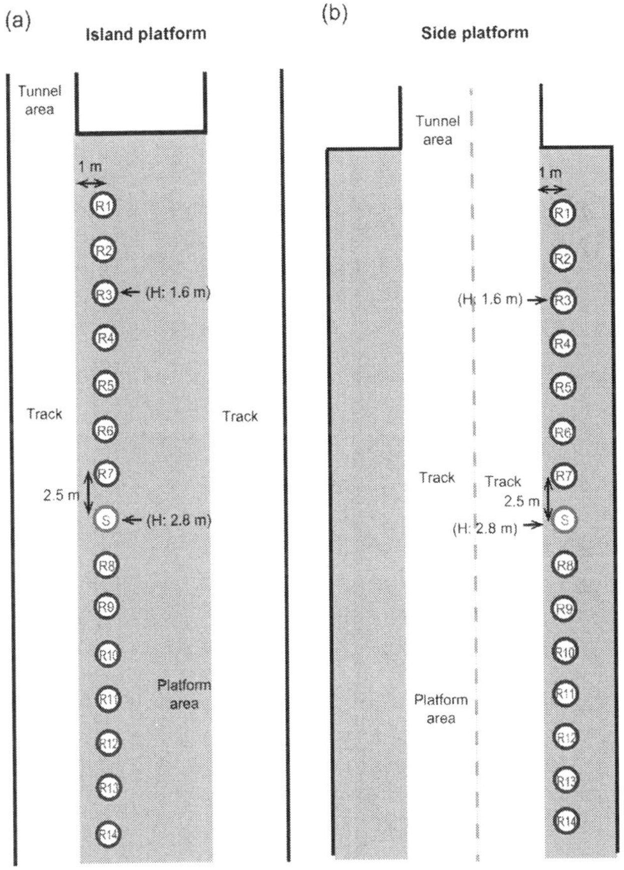

Figure 5. Source and receiver positions used to evaluate speech intelligibility of PA sounds (Configuration 1). (a) Island and (b) side platforms.

Figure 6 shows source and receiver positions of Configuration 2 for evaluation of the noise level of an approaching train to determine the noise reduction effects of the PSDs. Nine sound sources were located at a height of 0.5 m (0.02 m in the scale model) along the track with spacing of 15 m between them. Sound sources S1 to S3 were located inside the tunnel area. At the S1, S4 and S7 source positions, additional source heights of 2.3 m (0.092 m in the scale model) and 4.1 m (0.164 m in the scale model) were considered. Thirteen receivers were located at a height of 1.6 m with spacing of 5 m (0.2 m in the scale model) between them, in the same manner as the above configuration for sound field evaluation. R1 to R8 are classified as the front receivers, while R9 to R13 are classified as middle receivers.

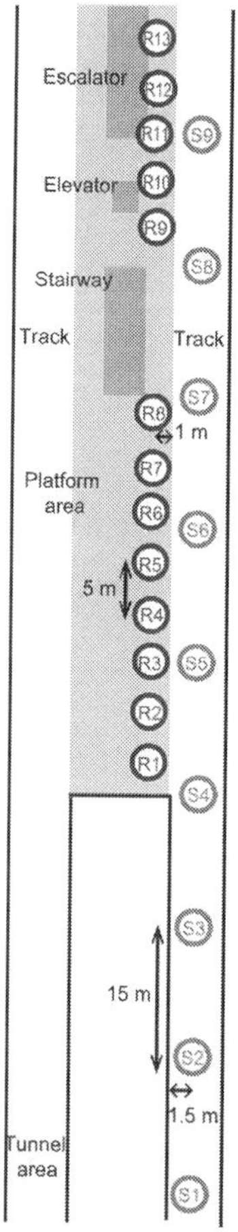

Figure 6. Source and receiver positions used to evaluate the noise level of an approaching train (Configuration 2).

2.6. Acoustic parameters

Four acoustic parameters, namely, the speech transmission index (STI), the sound pressure level (SPL), the reverberation time (RT, T30) and the inter-aural cross-correlation coefficient (IACC) [17, 18], were used to quantify the sound field characteristics using Configuration 1.

The STI is an objective measure of speech transmission quality. The STI measures certain physical characteristics of acoustic transmission in a room and is affected by the speech level, the frequency response of the room, the background noise level, the reverberation and other parameters. The STI ranges from 0 to 1. An STI of 1 indicates perfectly intelligible speech conditions. In general, the intelligibility rating is determined according to the STI value: 'excellent' for an STI of more than 0.75, 'good' for an STI of 0.6 to 0.75, 'fair' for an STI of 0.45 to 0.6, 'poor' for an STI of 0.3 to 0.45 and 'bad' for an STI of less than 0.3.

The SPL is a logarithmic measure of the sound strength at a specific position. To calculate the SPL, a reference value of 20 μPa and an A-weighted and octave band filters were commonly used in consideration of the human hearing threshold. Therefore, an SPL of 0 dB(A) indicates the minimum audible limit. The frequency range usually covers the range from at least 125 to 4000 Hz in octave bands.

The RT (T30) is defined as the time required for the SPL to decrease by 60 dB in a room, at a rate of decay that is given by a linear least-squares regression of the measured decay curve from a level 5 dB below the initial level to 35 dB below that level. The decay curve was obtained by reverse-time integration of the squared impulse response in each octave band. A higher RT value indicates a higher level of reverberation. To derive single number ratings, the SPL and the RT were averaged in the 500 and 1000 Hz octave bands.

The cross-correlation function between the signals obtained from the left and right ears is called the inter-aural cross-correlation function (IACF). The IACC is defined as the maximum absolute value of the IACF within the maximum possible inter-aural delay range for humans of 1 ms. The IACC correlates well with the subjective quality of 'spatial impression'. The spatial impression can be divided into two subclasses: Subclass 1 covers the broadening of the source, i.e. the apparent source width (ASW); Subclass 2 covers the state of diffusion of the reverberant sound field, i.e. the listener envelopment (LEV). To calculate the IACC, an A-weighted filter was included without spectral filtering. The IACC value ranges from 0 to 1. An IACC of 1 indicates that the signals from the left and right ears are identical, whereas an IACC of 0 indicates that they have no correlation at all. Usually, a lower IACC value indicates more diffused sound fields.

Also, the noise reduction level (NRL) overall bands was calculated as the SPL with the PSDs subtracted from the SPL without the PSDs to evaluate the noise level of an approaching train using Configuration 2.

2.7. Measurement setup in the scale model

Because the scale factor of 1/25 was used, a limited frequency range of up to 3840 Hz was measured through a tweeter loudspeaker (Clarion dome tweeter SRH294) and an analogue-to-digital/digital-to-analogue (AD/DA) converter (Roland Cakewalk UA-101) with a sampling rate of 192 kHz. Therefore, the STI in the scale model test was averaged from 500 to 1000 Hz. In addition, the IACC was not derived in the scale model test because of the use of a monaural microphone (B&K 1/4-inch microphone Type 4939-A-011, and B&K NEXUS conditioning amplifier Type 2690). During the measurements, the air temperature and the RH ranged from 21 to 26°C and 61 to 65%, respectively. The air absorption was corrected to calculate the RT as a real-scale condition at 20°C and 50% RH [19].

3. Results

3.1. Sound field characteristics by PSDs

The acoustic parameters were averaged for a single sound source and 14 receivers using Configuration 1. The parameters were expressed as relative values with reference to the NSD condition without any PSDs. Therefore, as an example, ΔSPL indicates the SPL without the PSDs subtracted from the SPL in each case.

3.1.1. Speech transmission index

Figure 7(a) shows the results for the STI values that were changed by the presence of the different types of PSDs from computer simulations and scale model testing. Apart from the case of the MCFH in the computer simulation, the STI was increased by the PSDs. In the MOFH case, the STI was maximally increased in both stations, by 3% in the island platform type and by 6% in the side platform type. The MHH cases showed similar STI increments to the MOFH cases. Because the fully closed cases (MCFH) showed the worst performance, it was found that the lower walls of the PSDs were important for improved speech intelligibility. The scale model results also confirmed the effectiveness of the PSDs in increasing the STI. However, the MCFH in the scale model showed increased STI values because the absorption properties of the model PSDs were slightly higher (0.07 in the mid-frequency range) than those of the real PSDs (0.03 in the mid-frequency range, tempered glass pane). In addition, the PSDs on the side platforms were more effective in increasing the STI than those on the island platforms, although the STI of the NSD case in the side platform-type station was 0.01 higher than that in the island platform-type station.

3.1.2. Sound pressure level

Changes in the SPL for the various types of PSDs from the computer simulations and scale model tests are shown in **Figure 7(b)**. In the computer simulations, the MCFH cases showed the highest increases in SPL of more than 3 dB for both island and side platforms. The MOFH cases showed relatively high SPL reinforcements. This tendency was also confirmed by the scale model results, although the SPL in the scale model was increased by approximately 1 dB because of the different absorption properties of the PSDs. However, the other cases showed only small changes in the SPL. In fact, the SPL was reduced in the FB case in particular.

Therefore, it was found that the upper walls of the PSDs were important for reinforcement of the SPL. Similar to the results for the STI, the side platform-type station was more effective in increasing the SPL than the island platform type.

3.1.3. Reverberation time

Figure 7(c) shows the results for the RT difference for the various types of PSDs from the computer simulations and scale model testing. In the computer simulations, the two full-height cases (MCFH and MOFH) showed greater RT reductions due to the PSDs in both the island and side platform cases. It seems that the reduction of the effective room volume by the PSDs was the main cause of the reduced RT. The MOFH cases in particular showed similar results to the MCFH cases, despite their upper walls being open. The scale model results confirmed

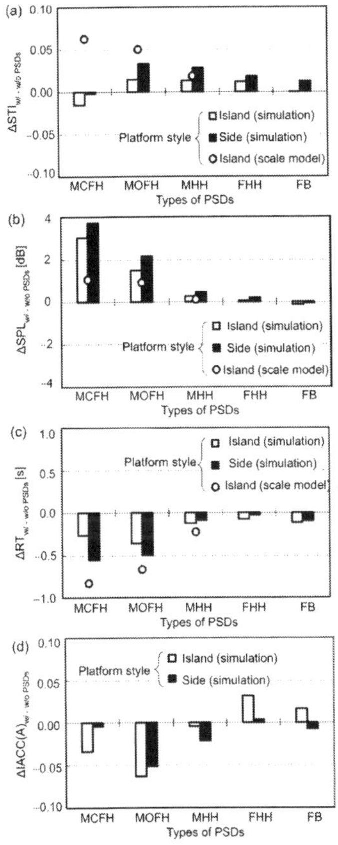

Figure 7. Differences in acoustical parameters between NSD and the other cases with PSDs. (a) STI, (b) SPL, (c) RT and (d) IACC.

that the RT was reduced by the PSDs. Also, the side platforms only showed greater RT reductions than the island platforms in the MCFH and MOFH cases. The FB also showed a reduction in RT when compared with the results of the SPL.

3.1.4. Inter-aural cross-correlation coefficient

The differences in the IACC for the various types of PSDs are plotted in **Figure 7(d)**. The MOFH showed the greatest IACC reduction for both station types. It seems that the coupling effects of the upper opening in the MOFH promoted the diffusion of reflections. The MCFH in the island platform type and the MHH in the side platform type also showed a reduction in the IACC values. However, the IACC increased in the FHH and FB cases on island platforms.

3.2. Noise reduction effects of PSDs

The NRLs were averaged from the nine sound sources and 13 receivers of Configuration 2. **Figure 8** shows the NRL results for the PSDs in the island and side platform configurations. The MOFH showed the highest NRL values for both island and side platforms. For the front receivers (R1 to R8), the NRL produced by the MOFH was 4.3 dB on the island platform and

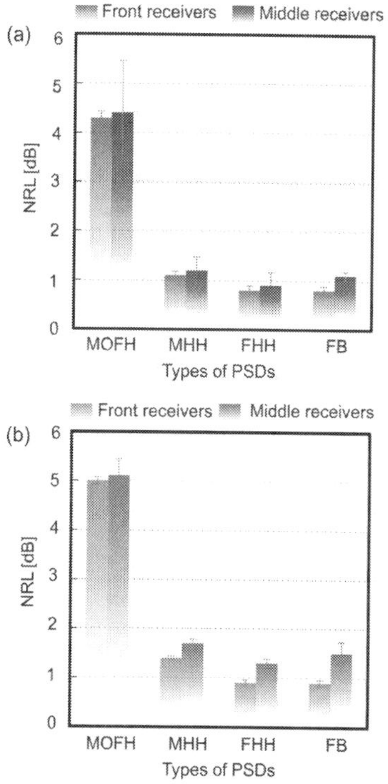

Figure 8. NRL values for each of the types of PSDs for (a) island and (b) side platforms.

5.0 dB on the side platform. These results show good agreement with those of the previous study that used field measurements [14]. The NRL produced by the MHH was 1.1 dB on the island platform and 1.4 dB on the side platform. The NRL values produced by the FHH and the FB were 0.8 dB on the island platform and 0.9 dB on the side platform. Therefore, the side platform-type stations showed higher NRL values than the island platform-type stations. This difference seems to be caused by the different boundary conditions of the platform sound fields when surrounded by the PSDs and the lateral walls.

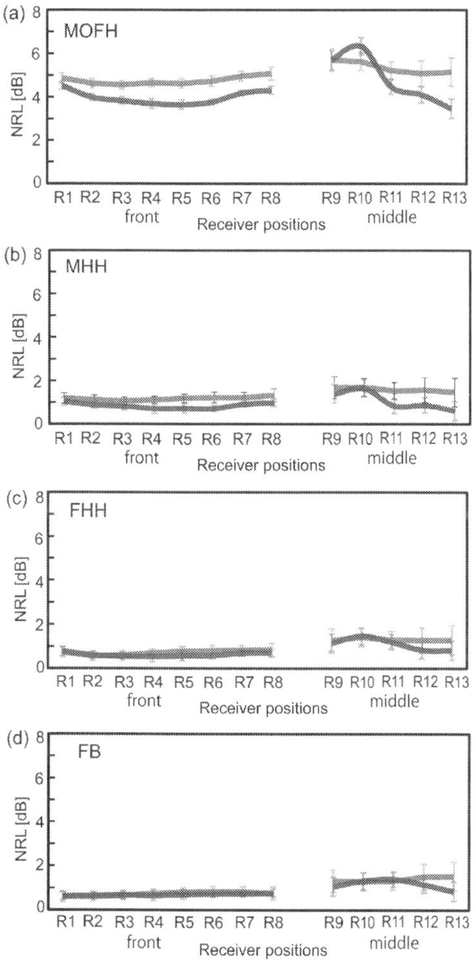

Figure 9. Variation of NRL values with receiver positions for (a) MOFH, (b) MHH, (c) FHH and (d) FB (blue line: island platform: red line: side platform).

In contrast, the middle receivers (R9 to R13) showed NRL values of 0.1–0.6 that were slightly higher than those of the front receivers. Smaller PSD profiles produced larger NRL differences between the front and middle receivers. The NRL seems to be affected by the presence of more diffusive interior elements in the middle platform area than in the front platform area.

3.2.1. *Effects of receiver positions*

To assess the effects of the receiver positions, the NRL distributions were plotted as shown in **Figure 9**. In the frontal platform area, stable NRL values were observed, with slight decay at the R4 and R5 positions. However, in the central platform area, dramatic changes in the NRL were observed. In particular, the R10 position beside the elevator shaft showed peak NRL values. The island platforms showed more fluctuating NRL values than the side platforms. These fluctuations seem to be caused by the opposite PSD walls in the island platform-type station, whereas the side platform-type station has only one PSD wall.

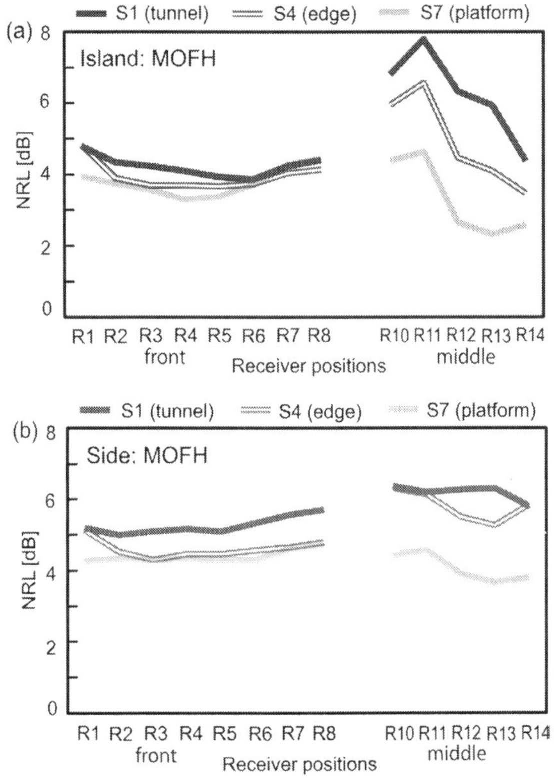

Figure 10. Variation NRL values according to source positions in case of MOFH for (a) side and (b) island platforms.

http://dx.doi.org/10.5772/64627

3.2.2. *Effects of source positions and heights*

Figure 10 shows the results for the NRL values according to the source positions in the MOFH case. Each NRL value for each receiver position was averaged for S1, S4 and S7. The MOFH showed higher NRL values for sound sources in the tunnel area than for the other source positions. This means that the PSDs are helpful in reducing relatively low level train noises. Particularly, large fluctuations in the NRL were observed in the middle platform area of the island platforms. Side platforms tend to show more relatively stable NRL values than island platforms.

Figure 11 shows the results for the NRL values based on the source heights in the MOFH case. Each of the NRL values for each receiver position were averaged for heights of 0.5, 2.3 and 4.1 m and were also averaged for S1, S4 and S7. On the island platform, the frontal receivers showed stable NRL values with respect to source height variation. However, the middle receivers on the island platform and the frontal receivers on the side platforms showed relatively large variation in their NRL values for different source heights. Lower sources tended to show higher NRL values.

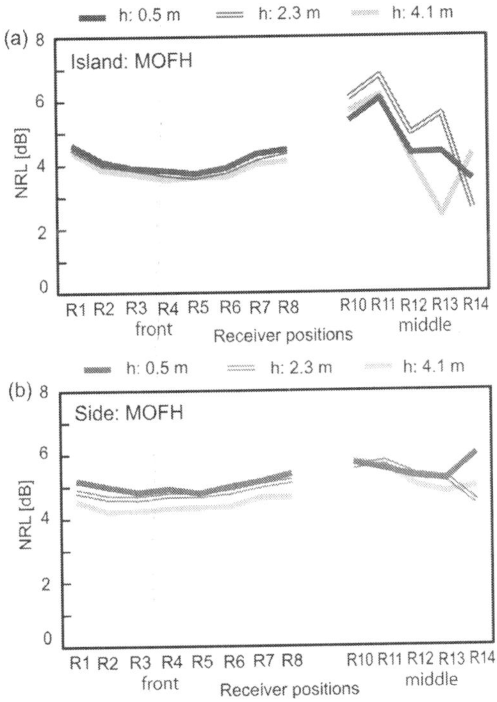

Figure 11. Variation of NRL values with source height in the MOFH case for (a) side and (b) island platforms.

4. Concluding remarks

In this study, the effects of PSDs on the sound field characteristics of underground stations and their noise reduction effects were investigated using both computer simulations and scale model testing. As a result, it was found that most types of PSDs were effective in increasing the STI with a higher SPL and reduced RT, apart from the MCFH type. Specifically, the MOFH was found to be the most effective type of PSD in terms of maximizing the STI with the lowest IACC. In addition, the noise reduction levels of these PSDs for train noise were derived for each type of PSD. PSDs on side platforms generally showed higher noise reduction performances than PSD on island platforms. In particular, the noise reduction level demonstrated by the MOFH type was 4.3 dB on island platforms and 5.0 dB on side platforms. Middle platform areas showed more unstable NRL values than frontal platform areas because they contained diffusive interior elements such as elevators or stairways. In conclusion, PSDs are helpful in reinforcing speech intelligibility and loudness for public address announcements while reducing train noise. As a potential future approach, the presence of background noise levels and the absorption effects of the passengers could also be considered to provide more realistic simulation results.

Acknowledgements

This work was supported by Grants-in-Aid for Scientific Research (B) (Grant No. 15H02771) and for Young Scientists (A) (Grant No. 23686086) from the Japan Society for the Promotion of Science.

Author details

Yoshiharu Soeta[1] and Yong Hee Kim[2*]

*Address all correspondence to: yhkim@kcl.re.kr

1 National Institute of Advanced Industrial Science and Technology (AIST), Ikeda, Osaka, Japan

2 Korea Conformity Laboratories (KCL), Ochang, Cheongju, Chungbuk, Korea

References

[1] Bhattachaya SK. Assessment of noise environment in a major railway station in India. Industrial Health. 1994; 32: 187–192.

[2] Endoh H, Suzuki H, Izumi Y, Hio Y. Appropriate sound level of broadcasting on station platforms under noisy environment. Railway Technical Research Institute Report. 2007; 21: 41–46 [in Japanese].

[3] Isumi Y, Fujii K, Iwase T. An investigation into the actual condition and subjective evaluation test of acoustical environment in railway station. Journal of Environmental Engineering. 2011; 76: 115–124 [in Japanese].

[4] Cohen A, Anticaglia J, Jones H. Sociocusis hearing loss from non-occupational noise exposure. Sound Vibration. 1970; 4: 12–20.

[5] Chang HC, Hermann ER. Acoustical study of a rapid train system. American Industrial Hygiene Association Journal. 1974; 35: 640–653.

[6] Johanning E, Wilder DG, Landrigan PJ, Pope MH. Whole-body vibration exposure in subway cars and review of adverse health effect. Journal of Occupational and Environmental Medicine. 1991; 33: 605–612.

[7] Shimokura R, Soeta Y. Evaluation of speech intelligibility of sound fields in underground stations. Acoustical Science and Technology. 2011; 32: 73–75.

[8] Kim SW, Seong KC, Kang BK. The study of introducing the screen door in subway station. Journal of Korea Institute of Healthcare Architecture. 2004; 10: 51–58 [in Korean].

[9] Oldfield A, Babic F. Acoustic design of transit stations. Journal of the Acoustical Society of America. 2012; 132: 2085.

[10] Lee MJ, Oh HW, Kim MJ. Measurement and analysis on the noise by train cars at platform of subway station. Journal of the Korean Society of Living Environmental System. 2009; 16: 126–133 [in Korean].

[11] Kim JC, Jeon SW, Koo DH. Analysis of noise reduction effect for platform screen door in the subway. Proceedings of Korean Society for Precision Engineering; 2009: 1067–1068 [in Korean].

[12] Nam JH, Park KS, Son WT, Ko JL, Shin JW. A study on the indoor noise reduction by installation of platform screen doors in a subway. Proceedings of the Society of Air-conditioning and Refrigerating Engineers of Korea; 2010: 1130–1135 [in Korean].

[13] Lee CM, Jung JG, Jung JS. Investigation and evaluation of noise level of the Busan subway. Journal of Korean Society of Environmental Engineers. 2011; 33: 243–250 [in Korean].

[14] Soeta Y, Shimokura R. Change of acoustic characteristics caused by platform screen doors in train stations. Applied Acoustics. 2012; 73: 535–542.

[15] Shimokura R, Soeta Y. Characteristics of train noise in ground and underground stations with side and island platforms. Journal of Sound and Vibration. 2011; 330: 1621–1633.

[16] Kim YH, Soeta Y. Design of diffusive surfaces for improving sound quality of underground stations. Proceeding of the International Symposium on Room Acoustics (ISRA), Toronto; 2013: P111.

[17] IEC 60268-16 (2003). Sound system equipment—Part 16: Objective rating of speech intelligibility by speech transmission index.

[18] ISO 3382-1 (2009). Acoustics–Measurement of room acoustic parameters—Part 1: Performance spaces.

[19] ISO 9613-1 (1993). Acoustics—Attenuation of sound during propagation outdoors—Part 1: Calculation of the absorption of sound by the atmosphere.

Chapter 8

Noise Control by Suppression of Gas Pulsation in Screw Compressors

Elvedin Mujić, Ahmed Kovačević, Nikola Stošić and Ian K. Smith

Additional information is available at the end of the chapter

http://dx.doi.org/10.5772/64795

Abstract

The various sources of noise in screw compressors have been determined, the most significant of which are gas pulsations and these have been analysed extensively in this chapter. The parameters most affecting them have been identified and different simulation tools have been used to quantify their effect, together with a brief overview of the capabilities of each of them. Resulting from these studies, methods of reducing the pulsations were identified and the improvements resulting from them were predicted. Tests were then carried out on an industrial screw compressor and good agreement was obtained between the predicted and measured levels of noise reduction.

Keywords: screw compressor, noise, gas pulsations, rotor rattling, transmission error

1. Introduction

Screw compressors are widely used in a number of applications, including refrigeration and air conditioning systems, transportation, the building industry, food processing and the pharmaceutical industry. Their ability to operate with a variety of working fluids and in the presence of injected liquids makes them a preferred choice to other types of compressor.

Over the past 30 years, these machines have been continuously improved, mainly by the introduction of better manufacturing methods, which enabled advanced rotor profiles to be developed. As a result, internal leakage and friction losses between the rotors have been reduced. These and many other improvements have increased their flow capacity, reliability and efficiency.

Screw compressors, however, generate a considerable level of noise during their operation, which sometimes inhibits the scope for their use. In addition, it is becoming increasingly difficult to satisfy new legislation that is focused on controlling noise at its source. More detailed investigation of the causes of noise, in these machines, is therefore necessary. Once these are identified and means of reducing their effect are found, attenuation procedures, such as insulation and absorption, are easier to apply and can reduce noise to even lower levels.

Apart from the direct benefits of noise reduction, there are also indirect benefits, such as in applications with tighter noise regulations, where less efficient types of machine can be replaced by screw compressors, which have higher reliability and lower maintenance costs, thereby bringing further benefits to both their operators and the environment.

Previous research activities have identified three different sources of noise in screw compressors, namely: mechanical noise, fluid flow sources and system vibration. Mechanical sources of noise have been thoroughly investigated by Stošić et al. [1] and Holmes [2] who proposed methods for reducing their effects. Investigations of gas pulsations, regarded as the most important fluid source of noise, have been published by different authors. These began in 1986 when Fujiwara and Sakurai [3] first measured gas pulsations, vibration and noise in a screw compressor. Subsequently, Koai and Soedel [4, 5], developed an acoustic model in which they analysed flow pulsations in a twin screw compressor and investigated their influence upon its performance. More recently, Sangfors [6, 7], Tanttari [8] and Huagen et al. [9] developed mathematical models for the prediction of gas pulsations in screw compressor suction and discharge chambers.

These authors explored the influence of various screw compressor parameters upon gas pulsations in the compressor suction and discharge chambers. The influence of the majority of them, which mainly affect the pressure difference between the compressor chambers, will be explained in more detail here. The investigation on gas pulsations carried by Mujić et al. [10] included the influence of the discharge port shape and area, as other important parameters, previously neglected.

In the present study, the effects of the different influential parameters were evaluated by the use of three different simulation models. A chamber model, which describes the screw compressor working cycle, was developed using MATLAB/SIMULINK. The model is very fast but not capable of predicting higher harmonics accurately. To improve on this, an existing full three-dimensional (3D) CFD model of flow within a screw compressor, and its discharge system, was applied. This model greatly improved the overall accuracy but was too time consuming. As a compromise a third model was developed that combined the best features of the first two. Predictions made with it were obtained in far less computational time with little difference from those obtained from the 3D CFD model, which agreed well with experimental measurements.

The results of the analyses showed that, although many parameters influence the level of gas pulsations, there are no practical means of altering most of them. However, changes in the shape of the compressor discharge port can have a strong effect and can reduce noise, even on

http://dx.doi.org/10.5772/64795

existing machines, as was confirmed by tests. For certain working conditions such alterations can cause a minor loss in performance, but, in other cases, they can even lead to improvements.

2. Working principle of screw compressor

Screw compressor operation is based on volumetric changes in all three dimensions. As shown in **Figure 1**, the compressor consists, essentially, of a pair of meshing helical lobed rotors contained in a casing. The flutes formed between the lobes on each rotor form a series of working chambers in which gas or vapour is contained. Beginning at the top and in front of the rotors, shown on the left, there is a starting point for each chamber where the trapped volume is initially zero. As rotation proceeds in the direction of the arrows, the volume of that chamber increases as the line of contact, between the rotor with convex lobes, known as the male rotor and the adjacent lobe of the female rotor, advances along the axis of the rotors towards the rear. On completion of one revolution, i.e. 360°of the male rotor, the volume of the chamber reaches its maximum and extends, in helical form, along virtually the entire length of the rotor. Further rotation then leads to re-engagement of the male lobe with the succeeding female lobe along a line of contact, starting at the bottom and front of the rotors and advancing to the rear, as shown on the right. Thus, the trapped volume starts to decrease. On completion of a further 360° of rotation by the male rotor, the trapped volume returns to zero.

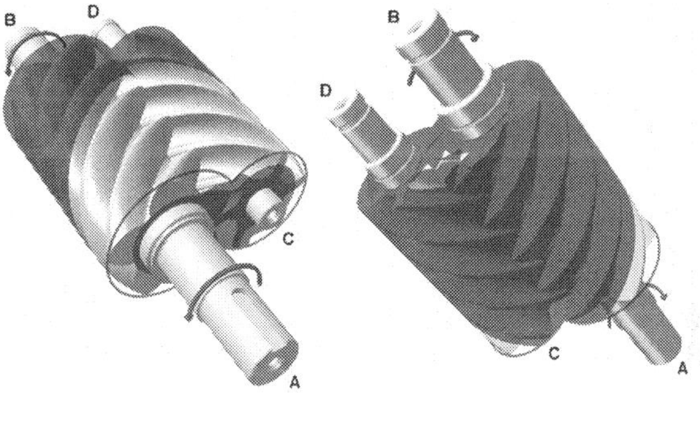

Figure 1. Working principle of a screw compressor.

The dark shaded portions show the enclosed region where the rotors are surrounded by the casing, which fits closely round them, while the light shaded areas show the regions of the rotors, which are exposed to external pressure. Thus, the large light shaded area on the left corresponds to the low pressure port close to shafts A and C while the small light shaded region between shafts B and D on the right corresponds to the high pressure port.

Exposure of the space between the rotor lobes to the suction port, as their front ends pass across it, allows the gas to fill the passages formed between them and the casing until the trapped volume is a maximum. Further rotation then leads to cut-off of the chamber from the port and progressive reduction in the trapped volume. This continues until the rear ends of the passages between the rotors are exposed to the high pressure discharge port. The gas is then expelled through this at approximately constant pressure as the trapped volume returns to zero.

3. Sources of noise within a screw compressor

During the compressor operating cycle, part of the energy is dissipated in the form of flow and mechanical disturbances such as gas pulsations or rotor rattling. These generate both system vibrations and pressure waves, with a range of frequencies and intensity levels, called noise. It is impossible to remove these noise sources completely, but some improvements are possible if efforts are directed towards the reduction of the disturbances created during the compressor working process. To do that, one has to first distinguish between the different mechanisms of noise generation and to classify them according to both their significance and the possibility of their reduction.

3.1. Mechanical sources of noise

The main source of mechanical noise is intermittent contact between the compressor rotors. This is the result of variation in the torque transferred from the male to the female rotor. New and more efficient rotor profiles introduce a very small negative torque to the female rotor compared with that of the male rotor. This torque, generated by the pressure-induced forces acting on the female rotor, is of the same order of magnitude as that created by other means, such as contact friction and oil drag forces. Since the negative female rotor torque acts in the opposite direction to the other two, the net torque may change in sign from negative to positive within one lobe rotation cycle as indicated by Stošić et al. [1]. This may cause instability in the female rotor rotational motion, resulting in flutter and, in the extreme case, rattling. Stošić et al. [1] proposed a new type of "silent" profile, which maintains positive torque on the female rotor and prevents this kind of noise from occurring.

Holmes [2] suggested another reason for noise generation, called transmission error. Transmission error occurs in the driven component of a screw pair when its instantaneous angular position differs from the theoretical angular position. This causes, earlier or later than expected, contact between rotor lobes and generates noise. Holmes suggested the reasons for the existence of the transmission error might be lead mismatch, lead non-linearity, pitch errors, housing bore imperfection, bearing deflections and rotor deflection due to the gas forces. Holmes proposed relieving the rotors' profiles to enable smoother contact between the lobes and reduce noise.

The noise reduction procedures proposed by both resulted in reported overall noise attenuation of 4–6 dBA.

3.2. Fluid sources of noise

According to Sangfors [6, 7], Koai and Soedel [4, 5], Tanttari [8] and Huagen et al. [9], gas pulsations are the main source of noise generated by fluid flow in screw compressors. These are created by unsteady fluid flow through the suction and the discharge ports which change the pressure within the suction and discharge chambers. The flow rate depends mainly on the pressure difference between the chambers and starts with the exposure and finishes with the cut-off of the suction or discharge port, as the rotors revolve past them.

A typical frequency spectrum of pressure pulsations in the suction and discharge chambers of the test screw compressor used in this investigation is shown in **Figure 2**. As can be seen, the gas pulsations are higher in the discharge chamber than in the suction chamber. However, while the discharge port is completely enclosed in its housing, the suction port may be more exposed to its surroundings, which are separated from the atmosphere only by the suction filter. Therefore, despite the smaller pulsations, noise generated in the suction chamber should require similar attention to that generated in the discharge chamber.

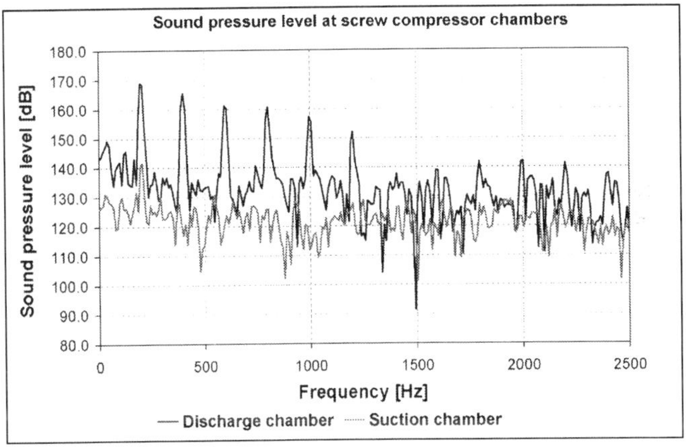

Figure 2. Sound pressure spectrum in suction and discharge chambers.

Another cause of fluid flow noise in screw compressors is turbulent fluid flow through the ports and the clearance gaps. Soedel [11] believes that turbulent noise contributes to the overall sound mainly in the higher frequency ranges between 3 and 6 kHz. However, compressor noise in that frequency spectrum is relatively low. So, this source of noise may be neglected.

4. Mathematical modelling of the discharge flow process

As mentioned in the introduction, three models for predicting flow properties within a screw compressor were applied to investigate the influence of the operational parameters and the compressor geometry upon gas pulsations. The first is a thermodynamic chamber model of the processes within a screw compressor, as described by Stošić et al. [12]. The second is a 3D CFD model which calculates the details of the fluid flow within a screw compressor, as given by Kovačević et al. [13]. As will be shown, these models are able to identify the influence of both operation and design parameters upon gas pulsations. However, the 1D model is insufficiently accurate while the 3D model requires excessive computational time.

A third, coupled simulation model was therefore employed to overcome the limitations of the first two. The new model provides a reasonably accurate result without the need for substantial computational time.

4.1. Mathematical model of screw compressor thermodynamics

Thermodynamic models of screw compressor performance [4, 6] have been proved to be useful tools for evaluating the influence of some of the design and operating parameters on gas pulsations in the discharge chamber. The one presented here is based on existing models described in [12, 14]. These were primarily developed to investigate the thermodynamic characteristics of a screw compressor within one working cycle. Starting from there, these models are able to calculate screw compressor integral parameters such as mass flow rate, compressor power, volumetric and adiabatic efficiency, etc. An updated model proposed by Mujić et al. [10] is able to account for changes in the geometry of the port shapes.

Since the mathematical basis of the model used here has already been well publicised, it will be presented in a very compact form, together with results obtained from it compared with those derived from experimental tests.

As shown in **Figure 3**, the thermodynamic model of a screw compressor is based on the assumption of five separate control volumes. The working chamber is periodically connected to the suction and discharge chambers through flow areas, which vary with time both in shape and size. The working chamber can also be connected to neighbouring chambers through clearances during the phases of the compressor cycle when the suction and discharge ports are closed. Other chambers are permanently connected to each other through openings of constant size throughout the working cycle.

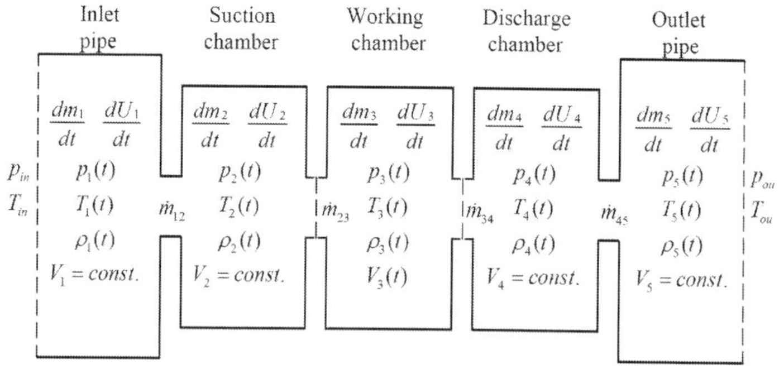

Figure 3. Thermodynamic simulation model of a screw compressor.

The model assumes that all thermodynamic values, such as pressure, temperature and density, are uniform within these control volumes. Any one of these control volumes can be considered as an open thermodynamic system, which exchanges fluid mass and energy with the environment, as shown in **Figure 4**. The mass and energy flowing in and out of any control volume affect the mass and energy level of the fluid trapped inside it.

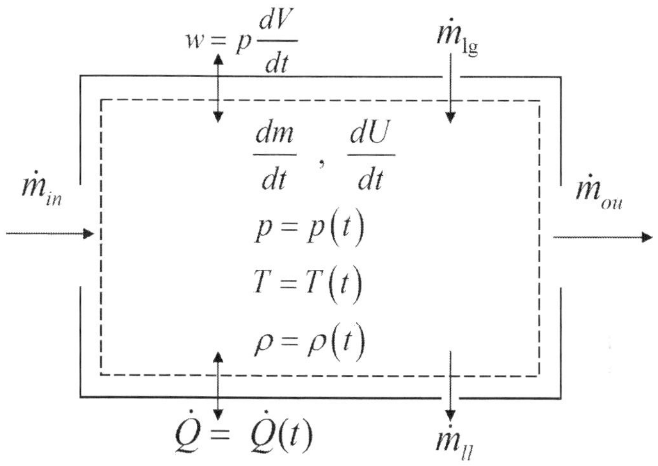

Figure 4. Screw compressor control volume for one-dimensional analysis.

The equation of mass conservation, which describes mass variation in the control volume, is given in Eq. (1):

$$\frac{dm}{dt} = \dot{m}_{in} - \dot{m}_{ou} \tag{1}$$

This equation is common for all the control volumes within the thermodynamic model. The mass inflow into the control volume consists of the suction flow from the previous control volume, the flow of injected fluid and the leakage flow which enters the control volume. The mass outflow from the control volume is obtained from its discharge flow and the leakage flow that leaves the chamber.

The equation for conservation of internal energy within the control volume may then be written as follows:

$$\frac{dU}{dt} = \dot{m}_{in}h_{in} - \dot{m}_{ou}h_{ou} + \dot{Q} - p\frac{dV}{dt} \tag{2}$$

The rate of change of internal energy within the control volume is equal to the difference of energy fluxes, which the control volume exchanges with its surroundings, together with the heat transfer through the control volume boundaries and the thermodynamic work.

Other phenomena within the control volume and at its boundaries are modelled by a number of algebraic equations, which describe leakage, inlet and outlet fluid velocities, oil injection and similar effects, and the differential kinematic relations which describe the instantaneous operating volume and how it changes with rotational angle or with time. The model also includes a thermodynamic equation of state of the fluid, required to complete and close the equation set.

4.1.1. Comparison of calculated and experimental results

The pressure history predicted by the thermodynamic model is shown for one of many sets of results in **Figure 5**. The predictions and measurements cover a set of points obtained for an industrial screw compressor at operational speeds in the 2000–6000 rpm range and discharge pressures in the range of 5–12 bar.

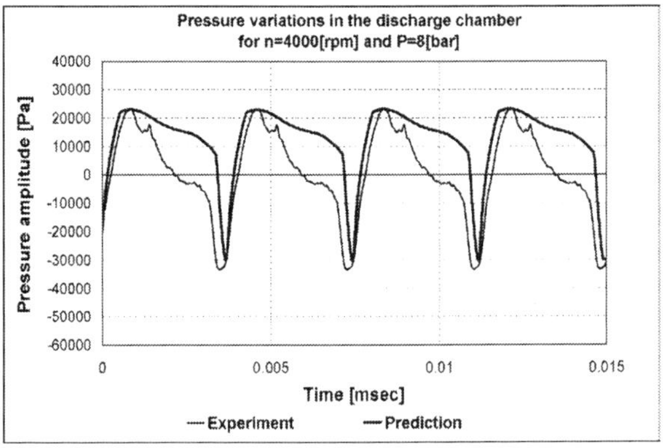

Figure 5. Gas pulsations calculated by 1D model, time domain.

These show that this thermodynamic model accounts very well for changes in the outlet pressure, which affect gas pulsations significantly. The model also takes into account variation of the shaft rotational speed and other compressor operational and geometrical parameters as, for example, changes in the discharge port geometry. The computation time is measured in seconds, when this model is employed. This makes it very useful for the analysis of the influence of any of the above-mentioned parameters on the level of the gas pulsations in the discharge chamber.

Although the results show that the predicted values of gas pulsations in the discharge chamber are in line with the amplitudes corresponding to the compressor fundamental frequency and its first harmonic, as is shown in **Figure 6**, this model does not predict higher harmonics accurately. This confirms that chamber, or even 1D, models are accurate only for a very narrow frequency range [5]. When averaged, for the amplitudes of the fundamental frequency and its first five harmonics, the prediction is from 40% to 65% in error. This value varies for different working conditions and increases for the higher harmonics.

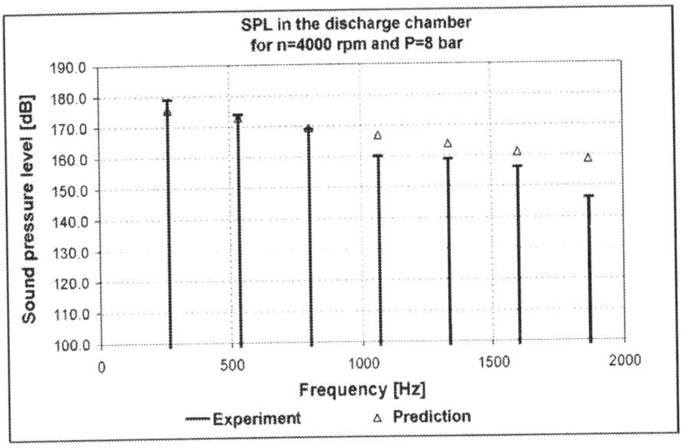

Figure 6. Sound pressure level, (SPL) of gas pulsations calculated by 1D model, frequency domain.

Koai and Soedel [5], who used a finite element method, showed that it is possible to achieve better agreement using a 3D model, which accounts for the complex geometry of the compressor discharge port and chamber.

4.2. Full 3D CFD model of a screw compressor

A 3D-CFD grid generation model, established by Kovačević et al. [13], was used to enable predictions of fluid flow in screw compressors to be made by means of commercial solvers Comet, StarCD or CFX. Pressure fluctuations in the compressor ports, including gas pulsations in the discharge chamber, were thus obtained, including fluid-solid interactions. Also, changes in the compressor geometry could be accounted for by means of a CAD interface.

Following recommendations expressed in [5], this model captured the higher harmonics more accurately.

4.2.1. Numerical grid of the screw compressor

To apply the calculation procedure, the compressor fluid domain is replaced by a numerical grid. The compressor fluid domain is divided into different fluid domains as shown in **Figure 7**. The grid consists of both moving and stationary parts.

The numerical mesh of the moving rotors is produced by a specially developed grid generator for screw compressor rotor domains, as described in more detail by Kovačević et al. [13]. The same grid generator may also be used for grid generation of stationary parts of the numerical grid, such as idealised suction and discharge domains. These stationary domains can also be mapped using a commercial grid generator, for example, the one included in Star CCM+, as shown in **Figure 7**. The generated rotor grid is fully structured while stationary chambers can generally be unstructured and may consist of tetrahedral, hexahedral or polyhedral grid elements, as shown in **Figure 7**. The relatively simple domains of the inlet and outlet pipes have been discretized with hexahedral grid elements. Once generated, the sub-domains are connected over coinciding boundary regions.

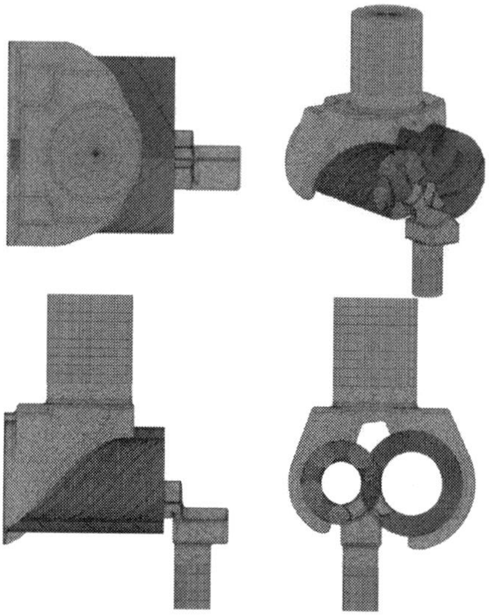

Figure 7. Numerical grid of screw compressor domains.

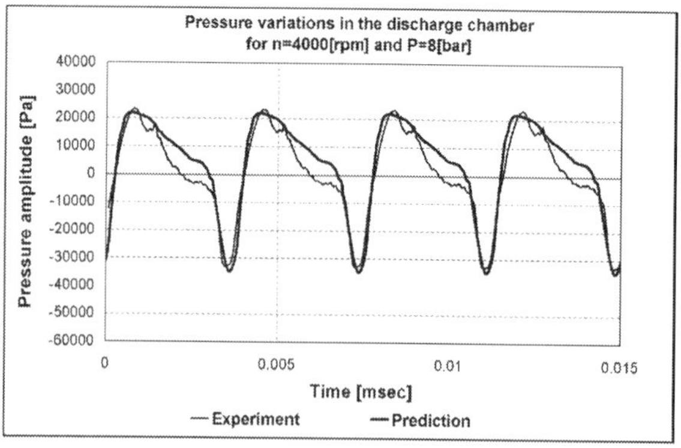

Figure 8. Gas pulsations calculated by 3D model, time domain.

4.2.2. *Comparison of calculated and measured results*

The numerical grid of the compressor fluid domain consisted of 652,017 grid elements. One discharge process was discretized through 50 time steps. Thus, according to the compressor speed, the size of the time step was in the range of 50–150 μs. The convergence criterion inside one time step was satisfied when residuals dropped by three orders of magnitude.

The gas pulsations predicted by the 3D model are shown in **Figure 8**, compared with the same set of measurements used for comparison with the thermodynamic model.

As expected, the results presented in the frequency spectra, in **Figure 9**, show an improvement over those obtained from the thermodynamic model presented in **Figure 6**. This improvement is mainly due to enhanced prediction of the higher harmonics by including the pressure history within the discharge chamber, while taking account of the discharge chamber geometry.

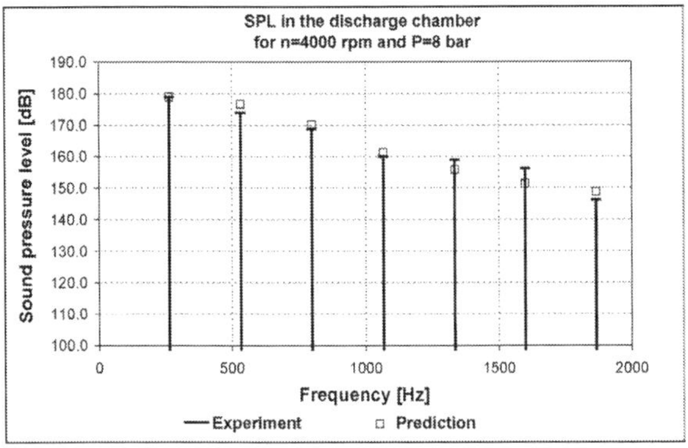

Figure 9. Gas pulsations calculated by 3D model, frequency domain.

The 3D model estimates non-uniform distribution of the pressure and other flow properties across the control volume, which creates pressure waves that affect the measurements.

Although the 3D model provided more accurate results, the time required for calculation was much longer than that for the thermodynamic model. The calculation of this case required between 24 and 30 hours on a PC with a 2.8 GHz Intel Xeon Dual Core processor and 2.5 GB RAM. Therefore, the evaluation of a number of different influential parameters could take weeks or even months. A new simulation model was therefore proposed to preserve the accuracy of the 3D model while reducing the calculation time.

4.3. Coupled model

The use of coupled models which combine a thermodynamic chamber model with a 3D model has been proposed to estimate pressure and flow changes in IC engine systems [15]. In such models, components of the system of secondary concern are simulated with a thermodynamic model and coupled with full 3D simulations for components of greater interest.

These coupled models combine the advantages of the fast computation and high flexibility of the thermodynamic model, with the enhanced capabilities of the 3D model. This permits more extensive investigation of the flow field, multiphase flow computation and chemical reactions within the cylinders.

A similar approach can be used to evaluate how the compressor operational parameters and its geometry influence gas pulsations. Therefore, a new coupled model was developed to take advantage of the 3D model while, at the same time, reducing computational time [16].

The main objects of interest were the geometrical shapes of the discharge chamber and the discharge port, both of which are in the fluid domain within the discharge chamber. The gas flow through them was therefore simulated by a 3D model. However, the other domains of the compressor, which were of secondary interest for this purpose, were analysed with sufficient accuracy by the thermodynamic model, already described in Section 4.1. The structure of the coupled model is shown in **Figure 10**.

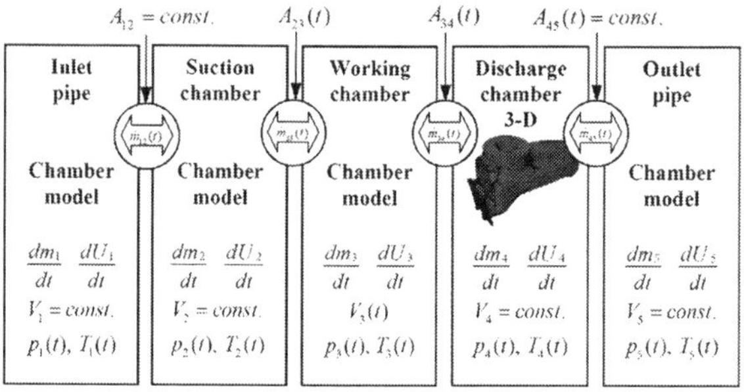

Figure 10. Structure of the coupled model.

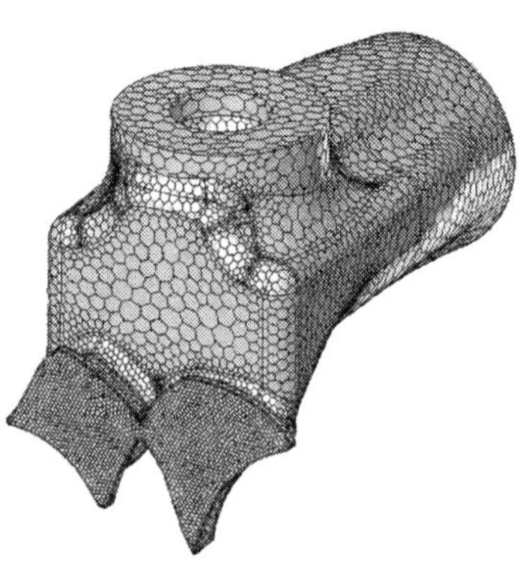

Figure 11. Numerical grid of the discharge chamber.

http://dx.doi.org/10.5772/64795

The 3D CFD model of the discharge chamber is simpler than the previously described 3D model because it does not contain any moving parts. This excludes the equation of space conservation from the calculation. The numerical mesh of the discharge chambers is shown in **Figure 11**.

A commercial CFD solver Star CCM+ was used to estimate the fluid flow in the discharge chamber. The thermodynamic model, used to simulate the remaining fluid domains, was programmed as a set of user sub-routines in the CFD code. The 1D model, defined, set the boundary conditions applied at the discharge port and the outlet of the discharge chamber for the 3D model as is shown in **Figure 12**.

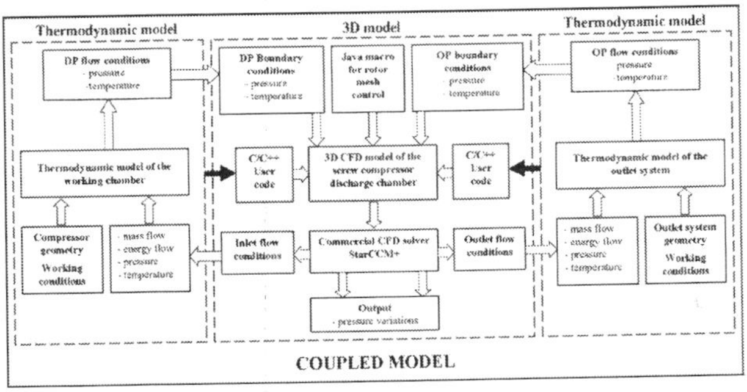

Figure 12. Coupling of thermodynamic chamber model and 3D model.

The arrangement shown in **Figure 12** enables a full two-way coupling of the models. The pressure and temperature calculated inside the working chamber and the outlet reservoir are used as the boundary conditions passed through the interface to the 3D model, where the flow field in the discharge chamber is calculated. The pressure and velocity values on the boundaries are then used to integrate the boundary mass flows, which are then used to calculate new values in the chamber model. This cycle is repeated until a converged solution is obtained, in which the mass and energy flows, pressures, velocities and densities are in balance. Once converged, the next time step is initiated with the new values of the chamber volume and the flow area and the procedure is repeated.

4.3.1. *Comparison of calculated and experimental results*

The numerical grid of the compressor fluid domain consisted of 86,000 grid elements. One discharge process was discretized through 90 time steps which represents a male rotor rotation of 1 degree per time step. Depending on the compressor speed, the length of the time step was in the range of 28–83 μs. The convergence criterion inside one time step was satisfied when the residuals dropped by three orders of magnitude. As for the full three-dimensional method, calculation of four cycles of the discharge process was necessary to obtain a convergent solution.

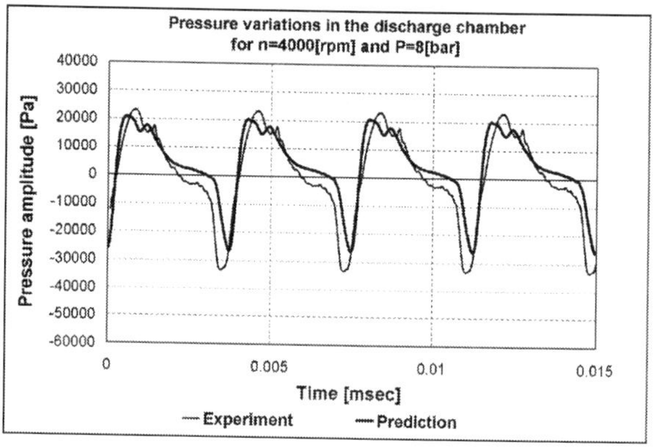

Figure 13. Gas pulsations calculated by coupled model, time domain.

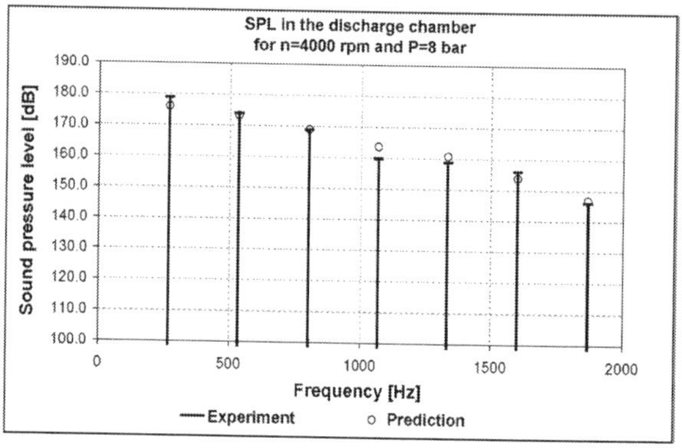

Figure 14. Gas pulsations calculated by coupled model, frequency domain.

The coupled model was validated by use of the same set of experimental data as for the thermodynamic and 3D models. Examples of the results calculated by the coupled model are shown in **Figures 13** and **14**. These results show far better prediction of the gas pulsations in the discharge chamber than those obtained from the thermodynamic model, at one position in the computational domain. As can be seen, these are almost identical with those obtained from the 3D model. The coupled model predicts pulsation amplitudes for the compressor fundamental frequency and higher harmonics reasonably well, as is shown in **Figure 14**.

Overall, the accuracy of the results calculated by the coupled model was considered to be sufficient to be used to analyse the influence of the compressor operational parameters and geometry upon the gas pulsations.

5. Parameters which influence gas pulsations

In order to reduce gas pulsation levels in the discharge chamber, the parameters which influence them have to be determined. Previous studies have reported the influence of some of them, as well as on other aspects of compressor design and operation. However, there was no explanation why these parameters were selected from the many others possible. To investigate this and to find if there are any other significant parameters, previously overlooked, the discharge process was analysed.

5.1. Main factors which affect gas pulsations

The working and discharge chambers, connected through the discharge port, are presented in **Figure 15**. Both of these are also connected to the rest of the system outside of the compressor. The compressor working chamber is connected to the other working chambers, or to the suction chamber, through leakage paths, while the discharge chamber itself is connected to the discharge pipe. Connection between the chambers and their surroundings is always present. However, the connection between the working and discharge chambers is of a periodic nature. Accordingly, the discharge process of the working chamber starts with the opening of the discharge port and lasts only until the port is closed.

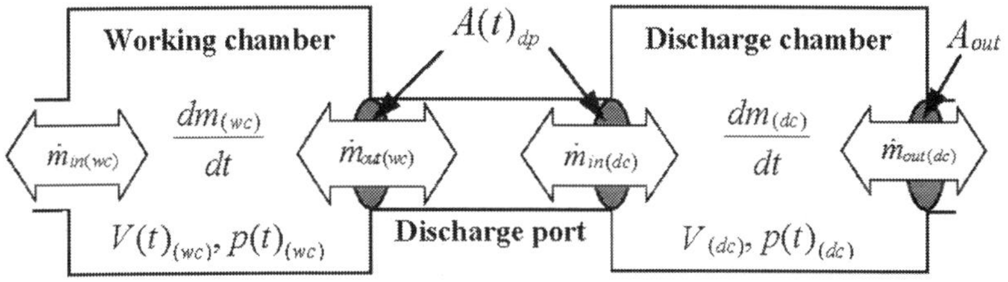

Figure 15. Screw compressor discharge system.

If the heat transfer through the chamber walls is neglected, then the gas condition in the chamber is determined by the change of the chamber volume and by the mass and energy transfer rates between the chamber and its surroundings. The volume of the working chamber is defined by the machine geometry and is expressed as a time-varying volume function, but the volume of the discharge chamber is constant. Therefore, it follows that the gas state in the discharge chamber is influenced only by the mass and energy transfer rates. This corresponds with the fact that gas pulsations in a chamber occur only if mass transfer rates to it are unsteady.

In **Figure 15**, two mass flows are shown between the discharge chamber and its surroundings. The first is the mass inflow into the discharge chamber from the compressor working chamber. The second is the mass outflow from the discharge chamber into the compressor discharge system consisting of pipes, oil separators and similar components. Both flows are time dependent and need to satisfy continuity Eqs. (3) and (4):

$$\dot{m}_{in(dc)} = \rho v A(t)_{dp} \tag{3}$$

$$\dot{m}_{ou(dc)} = \rho v A_{ou} \tag{4}$$

These equations show that both mass flows depend upon the instant gas density and the velocity. The fluid velocity is dependent upon the difference of fluid enthalpies in the chambers. For compressors, this is equivalent to the velocity being proportional to the square root of the pressure difference between the chambers. The gas density is also a parameter which is related to the chamber pressures. Therefore, the pressure difference between the working and discharge chambers is a parameter which influences the gas pulsation levels. Thus, the pressure difference between the discharge chamber and the pipe system is a result of the gas pulsations rather than their cause.

By analysing Eqs. (3) and (4), two more parameters can be identified which influence mass flow and later gas pulsations. The first is the outlet area, A_{ou} where the discharge chamber is connected to the discharge pipe. This outlet area is constant and it is good to make it as large as possible. A larger area will cause less pressure difference for the same flow between the discharge chamber and pipe and will therefore stabilise the pressure in the discharge chamber.

The second parameter, which influences the mass flow, which has not been properly considered in previous studies, is the cross-sectional area between the working and discharge chambers, $A(t)_{dp}$, defined at the discharge port. According to Eq. (3), the effect of variation of the discharge port size is of the same order of magnitude as those of changes in density or velocity. This implies that the gas flow variation and pressure pulsation can be altered by modifying the discharge port area function. This can be achieved by changing the shape of the discharge port.

The influence of different discharge ports has been noticed but not explained in papers published by Errol and Ahmet [17] and Mujić et al. [18]. The difference in noise, identified through their research, can be explained only by the different cross-sectional area of the ports. The authors are not aware of any published results of investigations of this phenomenon.

It follows that the two basic parameters which influence the level of gas pulsations in the discharge chamber are the pressure difference and the discharge port area function. In order to reduce the amplitude of gas pulsations in the discharge chamber, these two parameters need to be optimised. It is necessary to explore and to optimise at least one of these two parameters.

The influence of the pressure difference has been explored in many studies. The most relevant are probably the papers from Koai and Soedel [5] and Sangfors [9]. They recognise pressure difference as a cause for gas pulsations and they have explored the influence of the compressor operational and design parameters upon the gas pulsations.

5.2 Parameters related to working conditions

The discharge pressure is a parameter which will determine the pressure difference between the working and discharge chambers at the moment when the discharge process starts. The pressure difference is the smallest when the discharge pressure is equal to the pressure in the working chamber. This has been reported by Koai and Soedel [4]. They noticed that the gas pulsations are a function of the discharge pressure and have a minimum. This was also confirmed later by Sangfors [6]. According to Huagen et al. [9], this minimum corresponds to the discharge pressure that matches the machine built-in volume ratio. Koai and Soedel [4] claim that this minimum does not correspond exactly to that pressure, while Gavric and Badie-Cassagnet [19] consider that it occurs when there is slight under-compression.

5.2.1. Rotational speed

Sangfors [6] and Huagen et al. [9] concluded that the amplitude of the pressure pulsations during the discharge process increases with the rotational speed.

5.2.2. Oil influence

Oil has an attenuating influence upon the noise generation process. According to Sangfors [6], this is significant at harmonics higher than the third order, but Tanttari [8] states that this is noticeable only above the fifth order.

5.3. Compressor design parameters

5.3.1. Clearances

Reduction of the leakage flow within the compressor, due to smaller clearances, increases the noise level generated in the discharge port. Soedel [4] and Sangfors [6] reported that for the same working conditions, changes of compressor clearances alter the working chamber pressure and fluid flow through the discharge port.

5.3.2. Discharge chamber length

According to Sangfors [6], the gas pulsations and generated sound pressure level (SPL) are affected by the length of the discharge chamber. This influence is significant and the sound pressure level, being a function of the chamber length, has a minimum for a certain chamber length.

5.3.3. Number of rotor lobes

According to Sangfors [6], the number of rotor lobes influences the noise level. Rotors consisting of more lobes generally generate a lower sound pressure level in operation then those with fewer lobes.

All these parameters have a common factor in that they directly influence the pressure difference between the working and the discharge chambers. Operational parameters such as discharge pressure and compressor speed have a high influence on gas pulsations. However, those parameters depend on the purpose for which the compressor is used. This makes them inappropriate for optimisation. Compressor design parameters such as the built-in volume ratio, the clearances, the number of lobes and the sealing line length also affect gas pulsations. However, attempts to decrease gas pulsations by optimising any of these parameters, usually degrades the compressor performance and hence makes them unsuitable as parameters for optimisation.

Mujić et al. [10] showed how the time function of the discharge port affects gas pulsations in screw compressors. An example of varying the shape of the discharge port and its influence on gas pulsations will be presented here. For this purpose, models of the discharge process, already described, were used to determine the effect of the discharge port shape upon the level of gas pulsations. The other design parameters, such as the built-in volume ratio, clearances and sealing line length, were maintained constant. By this means any reduction in pulsations and change in efficiency can be attributed exclusively to the variation in port shape.

6. Gas pulsation reduction by use of discharge port alteration

Both experimental and numerical results were obtained for two different port shapes, to investigate their influence on the levels of gas pulsation in the discharge port. The numerical results, already given to evaluate the accuracy of the models, were used and will not be presented again. Instead, only the experimental results for different shapes of discharge port will be analysed.

6.1. Compressor testing range and instrumentation

For the purpose of this investigation an oil-flooded air screw compressor was used because it operates over a larger pressure range and develops higher gas pulsations than an oil-free compressor. In addition, an oil-injected screw compressor does not need synchronising gears, which affect the overall compressor noise level, and it may be driven at moderate speeds. Therefore, the noise generated by the test compressor was mainly caused by the gas pulsations. The compressor parameters are given in **Table 1**.

Compressor design parameters			**Main**	**Gate**
		Number of lobes	4	5
Centre distance	71 [mm]	Outer diameter	102 [mm]	80 [mm]
Rotor length	158 [mm]	Pitch diameter	63 [mm]	79 [mm]
Wrap angle	300 [deg]	Inner diameter	62 [mm]	40 [mm]
Pressure range	3–12 [bar]	Speed range	2000–6500 [rpm]	

Table 1. Screw compressor design parameters.

The test compressor was placed in a laboratory test rig, built to CAGI and PNEUROP standards. One set of measuring points covers the compressor speeds from 2000 to 6000 rpm at a discharge pressure of 8 bar, while the other set of points covers discharge pressures from 5 to 12 bar at a speed of 4000 rpm.

The discharge chamber pressure was measured by the use of an Endevco 8530C Piezoresistive pressure sensor to obtain a pressure function. The sound pressure level (SPL) around the test compressor was measured by pressure level indicators SJK. A sound Level Meter HML 323 was used to evaluate a correlation between the pressure function within the discharge chamber and the overall compressor noise. Measurements of the pressure, temperature, driving torque, air flow and compressor speed were also taken to estimate the compressor performance.

6.2. Modification of the discharge port

The original discharge port was modified to reduce the amplitudes of the gas pulsations by minimising any sudden flow between the working and discharge chambers at the beginning of the discharge process, when the pressure difference between the chambers is highest. According to Eq. (3), this flow is governed by the pressure difference and the size of the port area. Since the built-in volume ratio was not changed, the pressure difference was the same for both modified ports at the moment when discharge started. The original discharge port was designed to have the largest possible opening for any rotor position, in order to reduce flow losses. This was achieved by designing the port shape to correspond with the shape of the rotor trailing edges. Such an approach generates a discharge port area function with a high starting gradient, as shown by the light line in **Figure 16**. In that case the port has the largest opening when the pressure difference is largest.

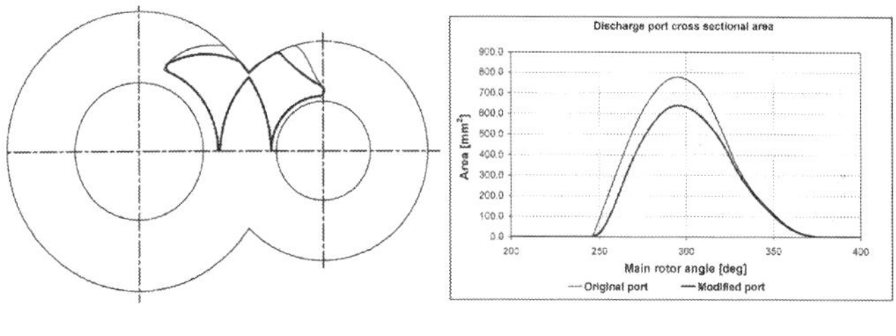

Figure 16. Shape and cross-sectional area of two different discharge ports.

To avoid this, a new port shape was devised, as shown by the bold line in **Figure 16**, where it can be seen clearly that only the port opening curves were changed. This altered the starting gradient of the port area function while maintaining the same built-in volume ratio as the original by keeping the remaining parts of the opening curves unchanged. Therefore, both ports opened at the same rotor position. The opening curves are simple in this second case and consist of only three arcs. They are therefore simpler than in the old port. The pressure function in the discharge chamber, predicted by the simulation models of the discharge process, showed a reduction in gas pulsations.

The shape modification of the port reduces the size of the port area. This difference increases at the beginning of opening and is highest when the port is fully open. It then decreases as the port closes and finally disappears when the leading edges of the following rotor lobes cover the opening curves of the port completely. Such a reduction must cause some flow losses.

6.3. Experimental verification of results

The shape of the compressor discharge port geometry, presented in **Figure 16**, was changed to verify the predicted results by locating metal inserts to form the new port. These inserts also covered the original opening curves on both sides of the port. These are illustrated in **Figure 17** by the two lines on the metal inserts. After modification, a new set of measurements was carried out in order to compare with the predicted gas pulsations.

The measured values of the pressure functions in the discharge chamber for the original and modified discharge ports are compared in **Figure 18**. These are shown for the original and modified port by light and bold lines, respectively. It can be seen that the gas pulsation amplitudes are reduced across the whole range of working conditions. As shown in **Figure 19**, the sound pressure level, generated by gas pulsations in the discharge chamber, is reduced.

More specifically, as can be seen in **Figure 19**, the overall noise in the compressor environment is attenuated by about 3 dB at 4000 rpm over the whole pressure range, while there is a noticeable noise reduction across the compressor speed range, varying from 2 dB at the lowest speed up to 5 dB at the maximum speed for an outlet pressure of 8 bar. The use of the model therefore represents a good step towards identifying the influential parameters and overall noise reduction

Figure 17. Modified shape of the discharge port.

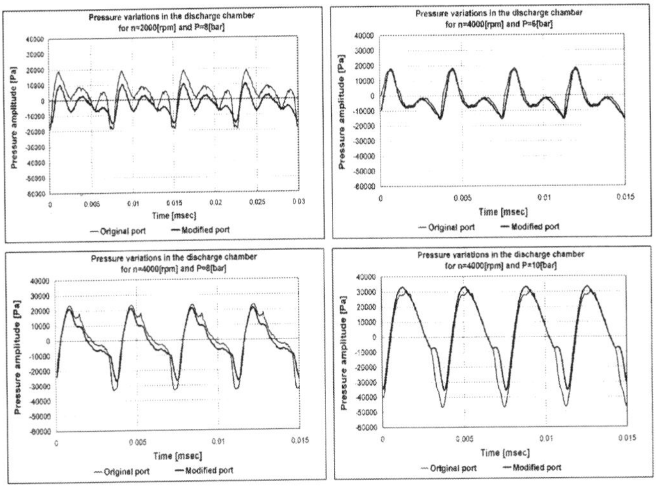

Figure 18. Comparison of experimental data for original and modified discharge ports.

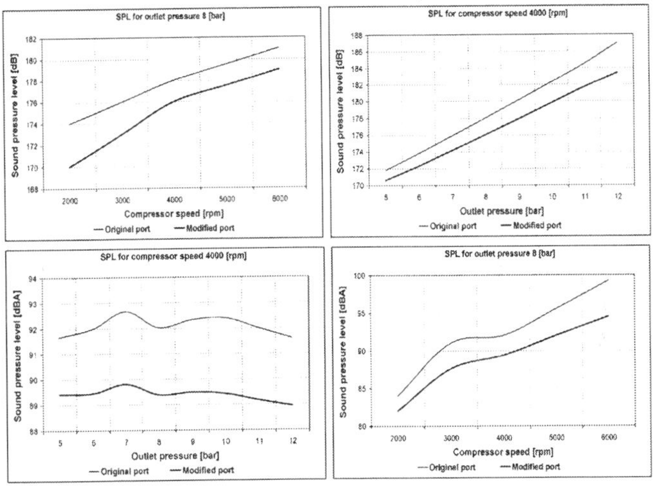

Figure 19. Comparison of calculated SPL inside the discharge chamber and measured SPL around compressor for original and modified discharge ports.

Since modification of the discharge port reduced the flow area, the flow losses were increased, thereby affecting the compressor performance. The compressor with the modified port requires more power than that with the original port and a comparison of the specific power for the two versions of the machine is presented in **Figures 20** and **21**. **Figure 20** shows that, over the whole speed range, the compressor with the modified port consumed more power at a constant outlet pressure of $p_{out} = 8$ bar.

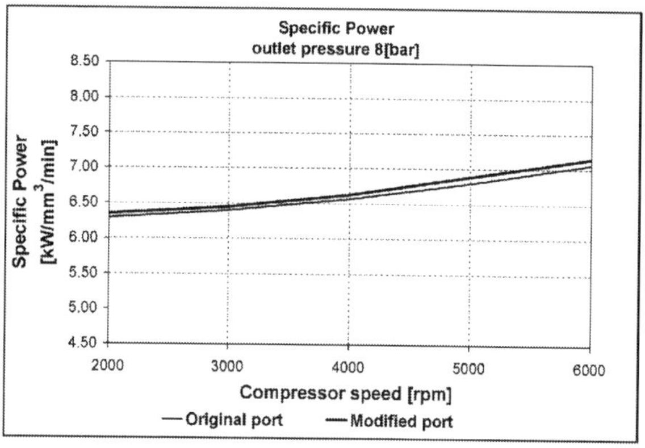

Figure 20. Compressor-specific power for different outlet pressures.

A comparison of the specific power, presented in **Figure 21** for a compressor constant speed of 4000 rpm, shows that the compressor with the modified port consumes more power for the majority of the pressure range. However, for the highest pressures in the range, the specific power is lower for the new port. The reason for this is that the new port shape reduces back flow when the compressor operates at higher pressures.

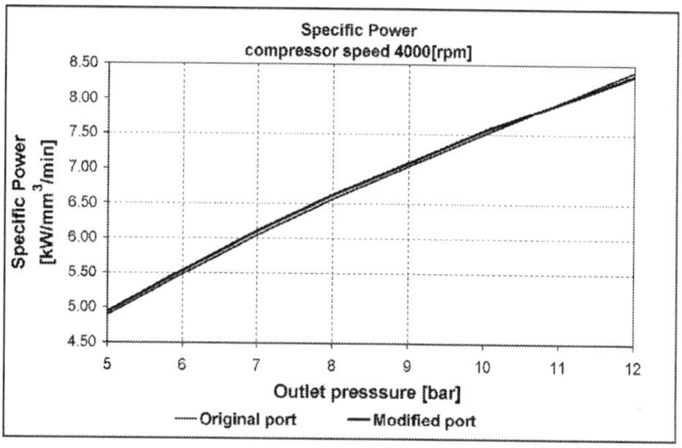

Figure 21. Compressor-specific power for different speeds.

http://dx.doi.org/10.5772/64795

7. Conclusion

Basic analysis has shown that the two most influential parameters affecting gas pulsations in a screw compressor discharge chamber are the pressure difference between the compressor working and discharge chambers and the discharge port area. Mathematical models applied to calculate the pressure function in a screw compressor discharge chamber have shown that the gas pulsations in a screw compressor discharge port and, consequently, the generated noise can be reduced by making appropriate changes to the shape of the discharge port. This has been confirmed by good agreement obtained over a wide range of speeds and pressures between predicted and measured values in a compressor with two different port shapes.

Nomenclature

A: area

h: specific enthalpy

m: mass

$\dot{m}$: mass flow

p: pressure

ρ: density

$\dot{Q}$: heat transfer

t: time

U: internal energy

V: fluid velocity

V: volume

Subscripts:

dc: discharge chamber

dp: discharge port

in: inlet

ou: outlet

Acknowledgements

The majority of the material presented in this chapter is reproduced from a PhD Thesis, "A Numerical and Experimental Investigation of Pulsation Induced Noise in Screw Compressors", by Mujić E, submitted at City University London in 2008, ref. [20].

Author details

Elvedin Mujić[1*], Ahmed Kovačević[2], Nikola Stošić[2] and Ian K. Smith[2]

*Address all correspondence to: elvedin.mujic@bitzer.de

1 Computational Engineering, Bitzer Kühlmachinenbau GmbH, Sindelfingen, Germany

2 Centre for Compressor Technology, City University, London, United Kingdom

References

[1] Stošić N, Mujić E, Kovačević A, Smith I K. Development of rotor profile for silent screw compressor operation. In: International Conference on Compressors and their Systems 2007; London, UK. London, UK: IMechE; 2007. p. 133–145.

[2] Holmes C S. Transmission error in screw compressors, and methods of THEIR? Compensation during rotor manufacture. In: International Conference on Compressors and their Systems 2005; London, UK. London, UK: IMechE; 2005. p. additional paper.

[3] Fujiwara A, Sakurai N. Experimental analysis of Screw Compressor Noise and Vibration. In: In The 1986 International Compressor Engineering Conference at Purdue; West Lafayette, US. West Lafayette, US:1986.

[4] Koai K L, Soedel W. Gas pulsations in twin screw compressors – Part I: Determination of port flow and interpretation of periodic volume source. In: The 1990 International Compressor Engineering Conference at Purdue; West Lafayette, US. West Lafayette, US: 1990. p. 369–377.

[5] Koai K L, Soedel W. Gas pulsations in twin screw compressors – Part II: Dynamics of discharge system and its interaction with port flow. In: In The 1990 International Compressor Engineering Conference at Purdue; West Lafayette, US. West Lafayette, US: 1990. p. 378–387.

[6] Sangfors B. Computer simulation of gas-flow noise from twin-screw compressors. In: International Conference on Compressor and Their Systems; London, UK. London, UK: IMechE; 1999. p. 707–716.

[7] Sangfors B. Modelling, measurement and analysis of gas-flow generated noise from twin-screw compressors. In: The 2000 International Compressor Engineering Conference at Purdue; West Lafayette, US. West Lafayette, US: 2000. p. 971–978.

[8] Tanttari J. On twin-screw compressor gas pulsation noise. In: The 29th International Congress and Exhibition on Noise Control Engineering; 27–30 August; Nice, France. 2000. p. 2369–2372.

[9] Huagen W, Ziwen X, Xueyuan P, Pengcheng S. Simulation of discharge pressure pulsation within twin screw compressors. Proceedings of the I MECH E Part A Journal of Power and Energy. 2004;218(4):257–264.

[10] Mujić E, Kovačević A, Stošić N, Smith I K. The influence of port shape on gas pulsations in screw compressor discharge chamber. Proceedings of the IMechE, Part E: Journal of Process Mechanical Engineering. 2008;222(E4):211–223.

[11] Soedel W. Sound and Vibration of Positive Displacement Compressors. New York, US: Taylor Francis Group; 2006.

[12] Stošić N, Smith I K, Kovačević A. Screw Compressors: Mathematical Modelling and Performance Calculation. Berlin/Heidelberg/New York: Springer; 2005.

[13] Kovačević A, Stošić N, Smith I K. Screw Compressors Three Dimensional Computational Fluid Dynamics and Solid Fluid Interaction. Berlin/Heidelberg/New York: Springer; 2006.

[14] Lee W S, Ma R H, Chen S L, Wu W F, Hsia H W. Numerical simulation and performance analysis of twin screw air compressor. International Journal of Rotating Machinery. 2001;7(1):65–78.

[15] Kolade B, Morel T, Kong S C. Coupled 1-D/3-D analysis of fuel injection and diesel engine combustion. Journal of Engines. 2004;113. p. 217–226.

[16] Kovačević A, Mujić E, Stošić N, Smith I K. An integrated model for the performance calculation of screw machines. In: International Conference on Compressors and Their Systems; London, UK. London, UK: IMechE; 2007. p. 757–765.

[17] Erol H, Ahmet G. The noise and vibration characteristics of reciprocating compressor: Effects of size and profile of discharge port. In: The 2000 International Compressor Engineering Conference at Purdue; West Lafayette, US. West Lafayette, US: 2000. p. 677–683.

[18] Mujić E, Kovačević A, Stošić N, Smith I K. Analysis and measurement of discharge port influence upon screw compressor noise. In: TMT 2005; Antalya, Turkey. 2005. p. 993–996.

[19] Gavric L, Badie-Cassagnet A. Measurement of gas pulsations in discharge and suction lines of refrigerant compressors. In: The 2000 International Compressor Engineering Conference at Purdue; West Lafayette, US. West Lafayette, US: 2000. p. 627–634.

[20] Mujić E. A Numerical and Experimental Investigation of Pulsation Induced Noise in Screw Compressors, PhD Thesis, City University London; 2008.

Chapter 9

Analysis of Acoustic Noise and its Suppression in Speech Recorded During Scanning in the Open-Air MRI

Jiří Přibil, Anna Přibilová and Ivan Frollo

Additional information is available at the end of the chapter

http://dx.doi.org/10.5772/64628

Abstract

The paper focuses on describing three methods of noise reduction in the speech signal recorded in an open-air magnetic resonance imager (MRI) working in a weak magnetic field during human phonation for the vocal tract modelling. This paper also analyses and compares spectral properties of the acoustic noise produced by mechanical vibration of the MRI device gradient coils. Then, the experiment with mapping of noise sound pressure level (SPL) in the MRI neighbourhood is described. The changes in acoustic noise spectral properties caused by loading of the holder of the lower gradient coils by the weight of the examined person lying in the scanning area of the MRI device is evaluated too. The influence of setting of the basic scan parameters of the used MR sequence (TR and TE times) on the spectral properties of the generated acoustic noise is also analysed. The results achieved are used to create a database of initial MR scan parameters such as the filter bank for noise signal pre-processing and to design a correction filter for noise suppression in the speech signal recorded simultaneously with three-dimensional (3D) human vocal tract scanning.

Keywords: acoustic noise, spectral analysis, noise suppression in speech, statistical analysis, magnetic resonance imager

1. Introduction

The non-invasive magnetic resonance imaging (MRI) technique enables scanning of the human vocal tract showing the configuration of the supraglottal resonant cavities (pharynx, oral and nasal cavities) during vowel phonation. The primary volume models of these acoustic spaces created from the MR images can then be transformed into the three-dimensional (3D) finite element (FE) models [1]. The FE models facilitate new possibilities in simulating and understanding of speech production. Synchronicity between image and audio acquisition must be ensured if the speech signal is recorded in parallel with 3D vocal tract scanning [2]. Further speech analysis and vocal tract modelling is conceivable only under the condition of an adequate signal-to-noise ratio as the noise is an inherent part of imaging by magnetic resonance [3]. The MRI device usually consists of three gradient coils to produce three orthogonal linear fields for spatial encoding of a scanned object. The noise is produced by these gradient coils due to rapidly changing Lorentz forces during fast switching inside the large static magnetic field [4]. It results in a significant mechanical vibration that subsequently propagates in the air as a progressive sound wave perceived by the human auditory system as a noise. Due to harmonically related audio frequencies of the produced acoustic noise, it can be analysed using similar methods as those used in speech signal processing.

There exist several approaches to reduce the acoustic noise produced during MRI scanning [5–7]. In our case, the tested person is lying and articulating in the scanning area of the MRI equipment while the scanning MR sequence is running to obtain an MR image of the human vocal tract. The recorded speech signal with the superimposed noise is then subject to offline analysis and processing. We have successfully used the noise reduction method based on the idea that the acoustic noise produced by the gradient coils of the MRI machine is a periodic signal and its fundamental frequency is represented by a typical peak in the real cepstrum in a similar way as it is for the voiced speech [8]. Our first proposed noise reduction method is based on the limitation of the real cepstrum of the noisy speech and clipping the "wrong peaks" corresponding to the harmonic frequencies of the acoustic noise. This method works well when the basic pitch period of the human voice differs from the repeating period of the running MR scan sequence. However, if both fundamental periods are the same, then alternative techniques must be selected. Our next tested approach to noise suppression uses subtraction between the short-time spectra of the audio signals recorded by two microphones [9]: the first one recorded the speech together with the acoustic noise and the second one recorded only the acoustic noise. Another method of speech enhancement in the MRI environment is based on spectral subtraction of the estimated periodic noise [10]. Therefore, we must know the statistical parameters of spectral properties of the noise produced by the MRI device. This third proposed method uses only one microphone picking up the noise of the running scan sequence that is followed by recording of the speech signal of phonation with the same noise background. As in the case of two-channel spectral subtraction, the natural logarithm and IFFT are applied on the resulting spectrum after subtraction, whereby the limited real cepstrum is obtained [11]. The final clean signal is reconstructed by the cepstral speech synthesizer [12].

This chapter is focused on comparison of the mentioned three methods of acoustic noise suppression in a speech signal recorded simultaneously with MRI scanning for 3D modelling of the human vocal tract. The main motivation of the work was to perform optimization of a correction filter to increase suppression of the acoustic noise while preserving spectral properties of the reconstructed speech and thus ensuring its good quality. The successfulness of different noise reduction methods can be evaluated subjectively—by visual comparison of spectrograms, spectral envelopes, etc. or with the help of objective approaches—numerical matching of determined differences between spectral envelopes, comparison of spectral distance values calculated between the obtained periodograms, etc. The special aspect of voice recording in the weak magnetic field environment is also discussed here. Next, the detailed analysis of basic and supplementary spectral features of the acoustic noise produced by the MRI device was carried out. The auxiliary measurement experiments also mentioned in this chapter include mapping of the acoustic noise pressure level in the MRI neighbourhood, analysis of the influence of different setting of scan sequence parameters as well as the influence of different body mass of the examined person situated in the scanning area on spectral properties of the generated noise signal.

2. Equipment and methods

The conventional whole-body MRI scanner with its specific acoustic noise makes speech recording in such an environment quite a challenging task that needs special solutions. One possible way of implementation is a construction of a special sound collector [13]; however, this realization requires a completely passive metal-free element without moving parts so that not to cause any artefacts in the scanning image due to induced magnetic field inhomogeneity. In practice, another solution was applied—speech signal collection by optical arrangements using special fibre optical microphones located in the scanning space [14]. In all cases, it allows for real-time processing of the speech signal simultaneously with obtaining of MR images. However, the practical implementation is significantly complicated (synchronization of both processes, special hardware for MRI device, etc.) and more expensive. In our case, when the tested open-air MRI device works with a weak magnetic field with the static magnetic induction up to 0.2 Tesla [15], an easier way to fulfil the condition about elimination of interaction with metal elements may be found. From this point of view, the most feasible solution for speech and noise recording is to use a condenser microphone located in the sufficient distance from the static magnetic field—see the overall view of the MRI equipment in **Figure 1a**.

In addition, due to the low basic magnetic field B_0 in the used MRI equipment, the scanning times are longer than those in the MRI devices working with 1.5 or 3 Tesla superconductive magnets. Prior to the start of each scan sequence, the quality factor (QF) parameter is automatically pre-calculated to obtain sufficient quality of a picture—see the obtained MR image of the human vocal tract in **Figure 1c**. The current setting of the scan parameters together with the chosen type of the sequence has also influence on the final scanning time. Typical scanning time is about 1 min, although for the 3D or Hi-Res scan sequences more than 2 min are necessary. However, phonation is performed in blocks with mean duration of 8, 15 or 30 s depending on physical condition of a tested person. The assumed physical position of the tested subject during the MRI session is typically the supine position, that is lying on the back with the neck placed inside the RF coil and between two gradient coils—see an arrangement photograph in **Figure 1b**. The previously performed experiments and measurements [8, 15] have shown that setting of the basic parameters of the scanning sequences—repetition time (RE) and echo time (TE)—has significant influence on the spectral properties of the acoustic noise produced by the MRI device. Values of these parameters result primarily from the chosen type of the scanning sequence as well as from other settings (field of view, slice thickness, number of sagittal images, etc.). The value of TR time affects mainly the basic noise frequency $F0_N$ in a similar manner as the vocal chords vibration affects the fundamental frequency $F0$ of the speech signal. On the other hand, setting of TE time, number of averages, etc. have effect on noise signal higher frequencies manifested as the changes of the first two formant positions (formant frequencies in a speech signal). From another experimental measurement and analysis follows that the mass of the examined person lying on the lower plastic holder of the gradient coil and the permanent magnet in the scanning area of the MRI device also changes the basic spectral properties of the generated acoustic noise [16].

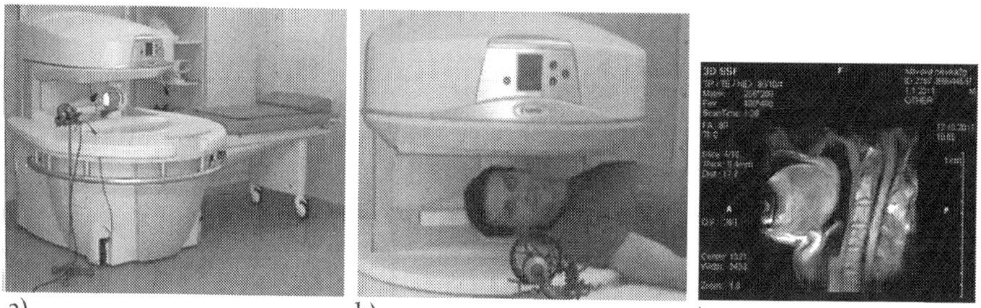

a) b) c)

Figure 1. Overall view of the open-air MR imager with the tested water phantom inserted in the RF coil inside the scanning area including an installed pick-up microphone (a), one microphone arrangement of the speech and noise recording experiment with the lying person (b) and obtained MR image of the human voice tract in a sagittal plane after 90° rotation obtained using the MR scan sequence 3D SSF (c).

2.1. Reduction in the gradient coil noise by cepstrum limitation and clipping

The first analysed approach of noise reduction in the recorded speech is based on clipping of the cepstrum peaks corresponding to the frequencies of the acoustic noise produced by the MRI gradient system. This method works in six steps—see the block diagram in **Figure 2**:

1. Calculation of the real cepstrum (including the energy) of the speech with the superimposed noise signal.

2. Detection of the pitch-period L_0 and the voiceness type of the speech signal.

3. Determination of the necessary minimum number of N_0 cepstral coefficients for sufficient log spectrum approximation.

4. Determination of the cepstral peaks corresponding to the frequencies of the acoustic noise in correspondence with estimated L_N value.

5. Limitation of the real cepstrum and cutting off the peaks.

6. Reconstruction of the cleaned speech signal by the pitch-synchronous cepstral speech synthesizer.

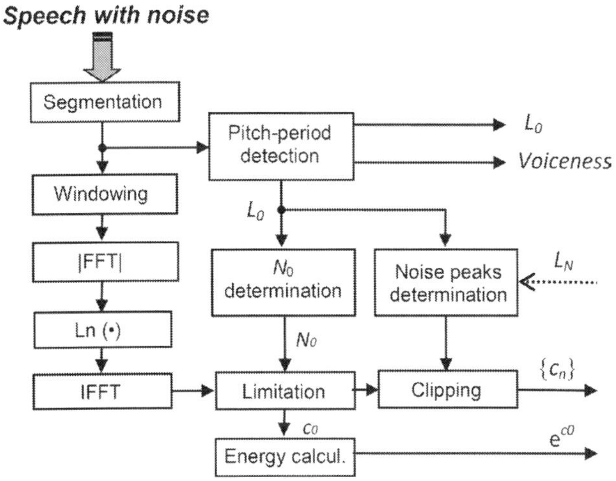

Figure 2. Block diagram of the noise reduction method based on clipping of the real cepstrum peaks.

The period L_N of the superimposed noise part of the analysed signal must be estimated externally, usually from the known repetition period (given by the currently chosen TR parameter) of the used MR scanning sequence.

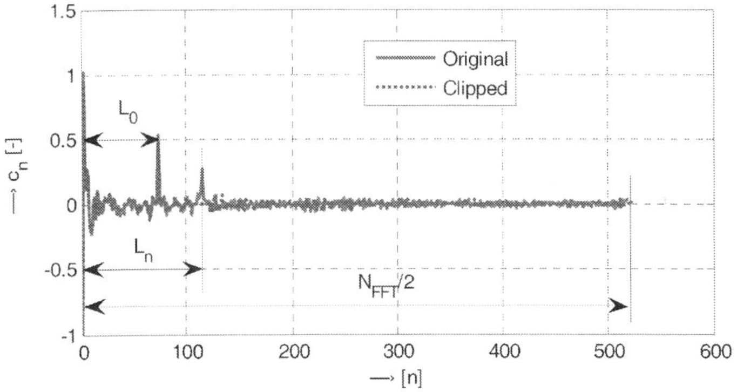

Figure 3. Graphic examples of noise reduction in the recorded speech by clipping and limitation of the real cepstrum; Lo = pitch-period of the speech signal (long vowel "a" articulated by a female person), Ln = period of the superimposed noise and N_{FFT} = 1024.

For cepstral analysis of the speech with the superimposed noise signal, the FFT of the Hamming-weighted signal frames is used for power spectrum computation and the natural logarithm followed by the inverse FFT is applied to get the symmetric real cepstrum—see an example in **Figure 3**. Limitation of the real cepstrum to its first N_0 + 1 coefficients can be described by the Z-transform as $C(z) = c_0 + c_1 z^{-1} + c_2 z^{-2} + \ldots + c_{N0} z^{-N_0}$. The truncated cepstrum represents an approximation of the log spectrum envelope

$$E(f) = c_0 + 2\sum_{n=1}^{N_0} c_n \cos(n \cdot 2\pi f) \tag{1}$$

where the first cepstral coefficient c_0 corresponds to the signal energy.

The reconstruction of the speech signal is realized by a digital filter implementing approximate inverse cepstral transformation. The system transfer function is defined as

$$G_F(z) = e^{c_0} \cdot \prod_{i=1}^{N_0} G_i(z) \tag{2}$$

and it can be implemented as a cascade connection of N_0 elementary filter structures. The system transfer function of the elementary filter is given by an exponential relation

$$G(z) = e^{\hat{S}(z)} \tag{3}$$

where the exponent is the Z-transform of the truncated speech cepstrum

$$\hat{S}(z) = \sum_{n=0}^{N_0} \hat{s}_n z^{-n} \tag{4}$$

and $\{\hat{s}_n\}$ represents the minimum phase approximation of the real cepstrum.

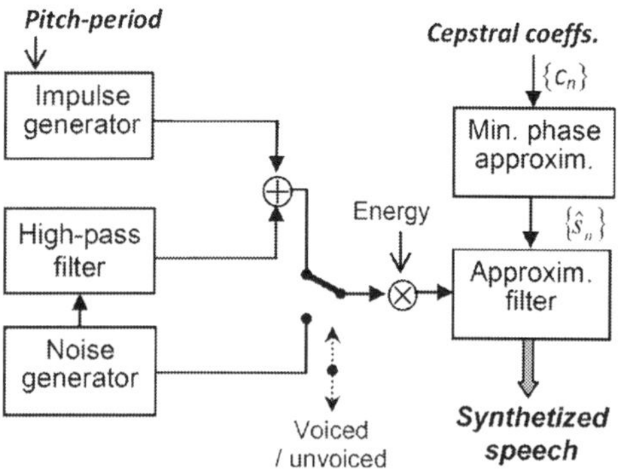

Figure 4. Block diagram of the cepstral speech synthesizer.

The transfer function of the vocal tract model is approximated by Padé approximation of the continued fraction expansion of the exponential function. For the voiced speech, the filter is excited by a combination of an impulse train and a high-pass filtered random noise, and for unvoiced speech, the excitation is formed by a random noise generator—see **Figure 4**. The error of the inverse cepstral approximation depends on the number and the values of the cepstral coefficients and the used approximation structure [12]. Still, the speech signal reconstruction must be carried out pitch synchronously to obtain correct proportionality between duration of voiced and unvoiced parts in the original and the resynthesized signal and to minimize the transient effect.

2.2. Two-channel noise reduction based on spectral subtraction

This method is based on the main idea that the noisy speech signal $x(n)$ is interpreted as an addition of a clean speech signal $q(n)$ and an additive noise $n(n)$.

$$x(n) = q(n) + n(n) \tag{5}$$

This noisy signal is segmented into the frames weighted by windows to obtain its short-time spectrum using the absolute value of the fast Fourier transform $|X(k)|$ of the signal $x(n)$

$$X(k) = \sum_{n=1}^{N_{FFT}} x(n) e^{-j\frac{2\pi}{N_{FFT}}nk} \tag{6}$$

where N_{FFT} represents the number of the points processed by FFT for calculation of the short-time spectrum $X(f, n)$. The enhanced speech spectrum can be obtained by subtracting the noise magnitude spectrum (recorded by the microphone Mic. 2) from the noisy speech magnitude spectrum (recorded by the first microphone Mic. 1)

$$S(f,n) = \left(|X(f,n)| - |N(f,n)|\right) \cdot e^{\phi_n(f,n)} \tag{7}$$

where $\varphi_n(f, n)$ represents the phase of the noisy speech spectrum and n is the index of the processed frame. After the subtraction, the natural logarithm and IFFT are applied on the resulting spectrum, whereby the limited real cepstrum is obtained. Finally, the clean signal is reconstructed by the cepstral speech synthesizer—see the block diagram in **Figure 5**.

2.3. One-channel noise reduction by spectral subtraction

The alternative proposed method uses only one microphone picking up the noise of the running scan sequence without phonation that is followed by recording of the speech signal during phonation with the noise produced by gradient coils of the MRI device with the same setting of the used MR scanning sequence.

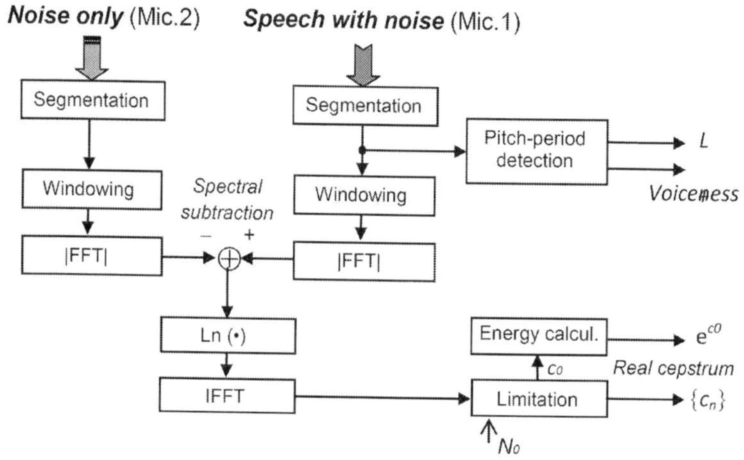

Figure 5. Block diagram of two-channel noise reduction by spectral subtraction and cepstral speech parameterization.

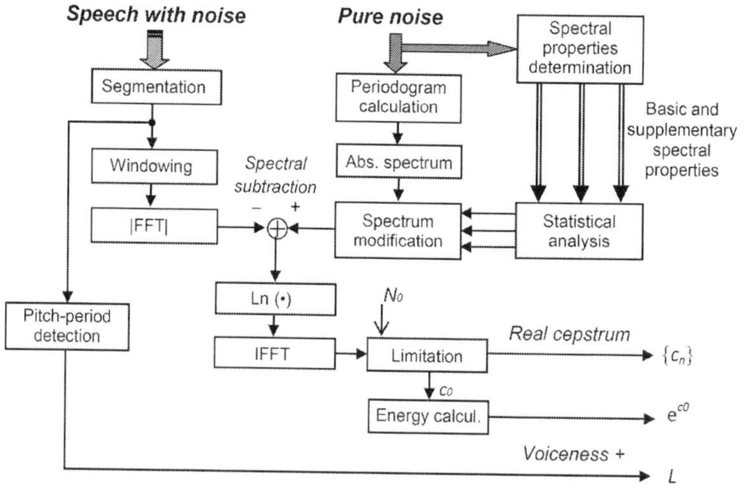

Figure 6. Block diagram of one-channel spectral subtraction method based on statistical analysis of noise spectral properties.

The algorithm contains practically the same processing of the speech signal with noise and the final cepstral analysis after spectral subtraction as it was in the previous two-channel method. However, totally different is processing of the pure noise signal part (typically shorter than the recoded speech signal of the long vowel articulation) is used to calculate the basic mean periodogram using the Welch method. After determination of the absolute spectrum $|S(k)|$, the basic spectral envelope is obtained. The pure noise signal is next segmentally analysed to determine the basic and supplementary spectral properties. The obtained spectral features are subsequently statistically processed, and the achieved values are used to modify the basic spectral envelope of the noise signal that is further subjected to spectral subtraction as described by the block diagram of the proposed method in **Figure 6**.

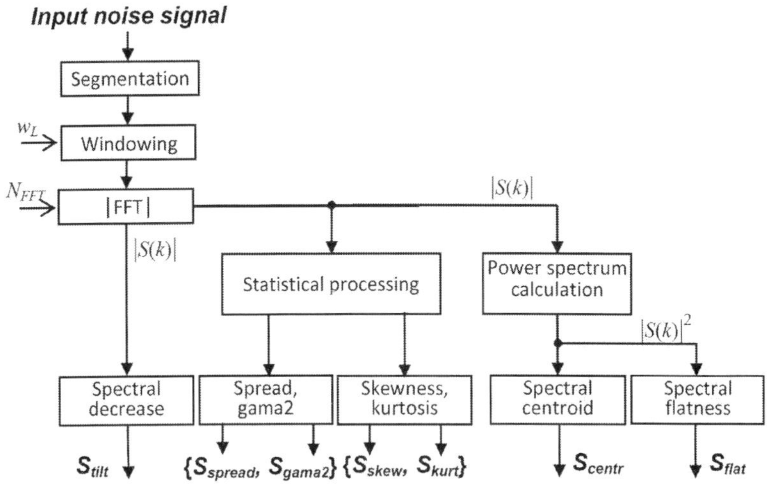

Figure 7. Block diagram of calculation and statistical processing of basic and complementary spectral properties of the noise signal.

The cepstral analysis (see the left part of the block diagram in **Figure 2**) can also be used for determination of basic and additional noise parameters. The basic as well as the supplementary spectral properties is usually determined from the frames (after segmentation and widowing —see the computation scheme in **Figure 7**). These noise spectral properties can be described with the help of the statistical parameters:

- The spectral spread (S_{spread}) parameter represents the dispersion of the power spectrum around its mean value

$$S_{\text{spread}} = E\left(x-\mu\right)^2 = \sigma^2 \tag{8}$$

where μ is the mean value or the first central moment and σ is the standard deviation (std) or the second moment of the spectrum values, and $E(t)$ represents the expected value of the quantity t in the processed data vector.

- The spectral skewness (S_{skew}) is a measure of the asymmetry of the data around the sample mean, and it can be determined as the third moment

$$S_{\text{skew}} = \frac{\mu^3}{\sigma^3} \tag{9}$$

- The spectral kurtosis (S_{kurt}) expressed by the fourth central moment represents a measure of peakedness of the shape of the spectrum relative to the normal distribution for which it is 3 (or 0 after subtraction of 3).

$$S_{\text{kurt}} = \frac{\mu^4}{\sigma^4} - 3 \tag{10}$$

- The measure S_{gama2} can be also applied to describe the statistical properties of the spectrum. It is defined using the spectral kurtosis and the spectral spread as follows

$$S_{\text{gama2}} = S_{\text{kurt}} / \left(S_{\text{spread}}\right)^2 \tag{11}$$

- The basic spectral properties include also the spectral decrease (tilt—S_{tilt}) representing the degree of fall of the power spectrum. It can be calculated by a linear regression using the mean square method.

The supplementary spectral properties describe the shape of the power spectrum $|S(k)|^2$ calculated from the noise signal:

- The spectral centroid (S_{centr}) is defined as a centre of gravity of the spectrum, that is the average frequency weighted by the values of the normalized energy of each frequency component in the spectrum. The S_{centr} in [Hz] can be calculated as

$$S_{\text{centr}} = \frac{\sum_{k=1}^{N_{FFT}/2} k\left|S(k)\right|^2}{\sum_{k=1}^{N_{FFT}/2} \left|S(k)\right|^2} \cdot \frac{f_s}{N_{FFT}} \tag{12}$$

where f_s is the sampling frequency.

- The spectral flatness (S_{flat}) determined as a ratio of the geometric and the arithmetic mean values of the power spectrum

$$S_{\text{flat}} = \frac{\left[\prod_{k=1}^{N_{FFT}} \left|S(k)\right|^2\right]^{\frac{2}{N_{FFT}}}}{\frac{2}{N_{FFT}} \sum_{k=1}^{N_{FFT}} \left|S(k)\right|^2} \tag{13}$$

can also be used for determination of a degree of periodicity in the signal. The S_{flat} values lie in the range of <0 ~ 1>: the zero value represents the signal that is totally periodical (e.g. a pure sinusoidal signal); in the case of $S_{flat} = 1$, the totally noisy signal is classified (e.g. the white noise signal).

2.4. Comparison of noise reduction methods by spectrograms and spectral envelopes

Several methods can be utilized for visual comparison of effectiveness of different approaches for noise reduction in the speech. A spectrogram is a typical representative graph for comparison in the time/frequency domain. The approach of visual comparison of the whole spectrograms has a disadvantage in the subjectivity (strong dependence on the person who makes the matching). To decrease this subjectivity, the detailed analysis of only the selected part–the nearest region of interest (ROI) area—is necessary. In this manner, more precise matching of differences can be obtained—see an example in **Figure 8**.

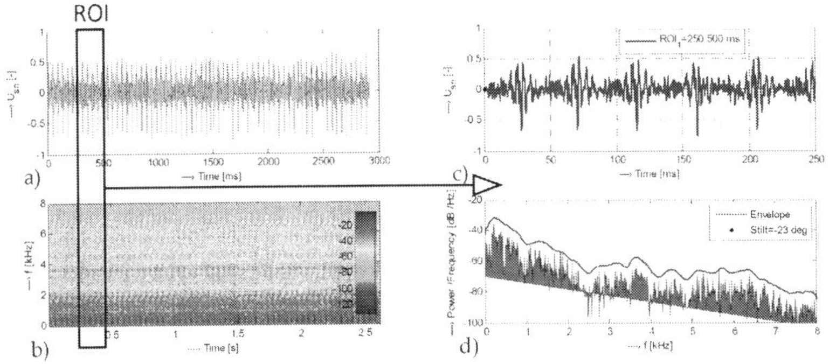

Figure 8. Example of speech and noise signal processing (f_s = 16 kHz): the whole recorded signal (a), the corresponding spectrogram (b), the selected ROI in 250-ms time interval (c) and the calculated spectral density together with its envelope and a tilt in degrees (d).

Generally, it can be said that the periodogram represents an estimate of the power spectral density of the input signal. When the N_{FFT}-point FFT algorithm is used for computing of the PSD as $S(e^{j\omega})/f_s$(with sampling frequency f_s in [Hz]), we can obtain the resulting spectral density in logarithmic scale expressed in [dB/Hz]. These PSD graphs can be used for visual comparison as well as for numerical matching with the help of the calculated spectral difference (ΔP) or using other additional spectral parameters (spectral tilt). To obtain the smoothed spectral envelope, the Welch mean periodogram in [dB] can be used, too, as documented by the right graph in **Figure 9d**. The resulting spectral envelopes can be used for subsequent comparison. For exact numerical comparison (objective matching method), it is possible to calculate the difference between the spectral envelopes, for example between the original noisy speech signal and the signal after noise suppression. It holds also in this case of comparison that evaluation in selected frequency sub-ranges must be applied for obtaining more precise results. Therefore, the comparison of spectral envelopes was divided to three ranges: the full frequency range up to $f_s/2$ (0–8k), the low-frequency sub-band up to 2.5 kHz (0–2k) and the middle frequency sub-band of 2~6 kHz (2–6k)–when the sampling frequency is f_s = 16 kHz. Then, the spectral distances D_{RMS} (by the RMS method) between these envelopes were calculated to compare the noise suppression methods for all three frequency ranges. In the case of the low-frequency band of 0–2k, the frequency F_{max} corresponding to the maximum difference ΔP_{max} was localized and determined, as documented by an example in **Figure 9**.

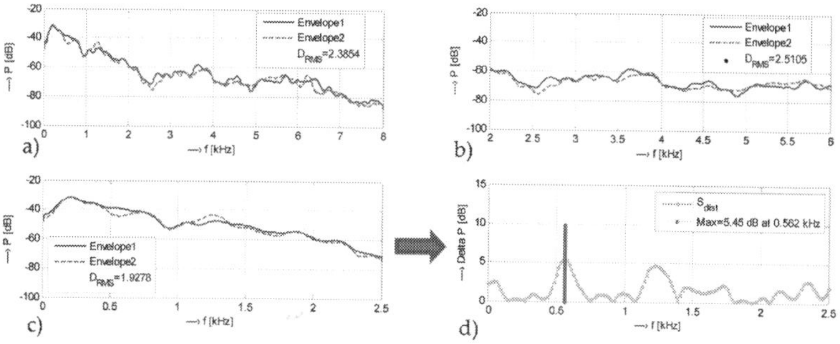

Figure 9. Example of visualization and calculation of differences between two spectral envelopes of the speech signals with the acoustic noise in the full frequency range 0~$f_s/2$ (a), in the middle frequency sub-band of 2~6 kHz (b), in the low-frequency band 0~2.5 kHz (c) and the spectral distance between envelopes in the low-frequency band (d).

3. Performed noise measurements and speech recording experiments

The open-air MRI equipment E-scan OPERA working with the weak magnetic field of 0.178 Tesla [17] was used for realization of all the experiments and measurements. This type of the MRI device includes also an adjustable bed, which can be positioned in the range from 0° (at the left corner near the temperature stabilizer) to 180°—see the overview photograph in **Figure 1a** and a principal angle diagram of the MRI scanning area in **Figure 10c**. The temperature stabilizer produces an additional acoustic noise, but its sound pressure level L_0 is almost constant so it can be easily subtracted as a background. As the noise depends on the position of the measuring microphone, the directional pattern of the noise source in the MRI scanning area must be measured first. On the basis of the obtained results, the design of the final arrangement of the recording experiment is carried out. It means that the optimal recording microphone position and parameters (the distance from the central point of the MRI scanning area, the direction angle, the working height and the type of the microphone pickup pattern) must be chosen. Therefore, the performed analysis of the acoustic noise produced by the gradient system of the MRI device consists of two phases. The first step is the auxiliary measurement of the acoustic noise distribution in the MRI scanning area and the mapping of the noise sound pressure level (SPL) in the MRI neighbourhood (see the arrangement photograph and principal angle diagram of the MRI scanning area in **Figure 10**):

- Directional pattern of the acoustic noise in the MRI neighbourhood in the angle intervals of 12.5° at the distances of {45, 60 and 75 cm} from the central point of the scanning area, at the height of 85 cm from the floor.
- Directional pattern of the acoustic noise in the angle intervals of 12.5° at the distance of 60 cm in three measurement heights of {75, 85 and 95 cm} from the floor corresponding to the sound level meter location on the level of the bottom gradient coil, in the middle between both coils and finally on the level of the upper coil.
- Noise SPLs at the directions {30°, 90° and 150°} at the distance of 60 cm in the height of 85 cm with different male/female testing persons lying in the scanning area of the MRI device without phonation; comparison with values obtained by the measurement using the water phantom.

a) b) c)

Figure 10. Arrangement photograph of the SPL distribution measurement in the MRI Opera neighbourhood with the sound level meter at the position of 30° in the 45 cm distance from the middle point of the scanner (a), detailed photograph of the MRI scanning area with a testing water phantom inserted in the RF coil (b) and a principal angle diagram of the MRI scanning area (c).

The sound level meter of the multi-function environment meter Lafayette DT 8820 (with the range set to 35–100 dB) was used for the noise SPL measurement and mapping. In the first two experiments, the spherical phantom weighing about 0.75 kg filled with doped water solution was placed in the RF knee coil (see the photograph in **Figure 10b**), and the measurement was carried out during the MR scan sequence SSF 3D execution (the patient bed was at the position of 180).

Afterwards, three basic types of noise and speech signal recordings were realized:

1. The noise produced by the gradient coils recorded by one pick-up microphone only—the tested person in the scanning area of the MRI device without phonation—the Mic. 1 located at 150.
2. The signals recording by the two pick-up microphones during execution of a scan MR sequence and phonation of the person lying in the scanning area of the MRI device (the noise and speech by Mic. 1 located at 90 and the noise only by Mic. 2 at 180 behind the lying person to minimize crosstalk of the speech signal).
3. The phonation signal with an additive noise recording by one pick-up microphone (Mic. 1 at 90—see the middle photograph in **Figure 1b**).

The demonstration examples of typical speech and noise waveforms for all three above-mentioned noise reduction methods are shown in **Figure 11**.

The electrical signals from the pick-up 1 'Behringer dual diaphragm condenser microphone B-2 PRO (as Mic. 1) and the RØDE NTK 1' condenser microphone (Mic. 2–in the case of two-way noise reduction method, both with a cardioid directional pattern) were recorded by means of the Behringer XENYX 502 analogue mixing console connected to a notebook by the USB interface UCA 202. Synchronization of the noise and/or speech recording and the MR scan process was realized manually by the console operator. The noise as well as speech signals was originally recorded at 32 kHz and then resampled to 16 kHz. A typical duration of the recorded signal was about 30 s, and for further signal processing, the stationary parts lasting 15 s were selected using the sound editor program Sound Forge 8.0.

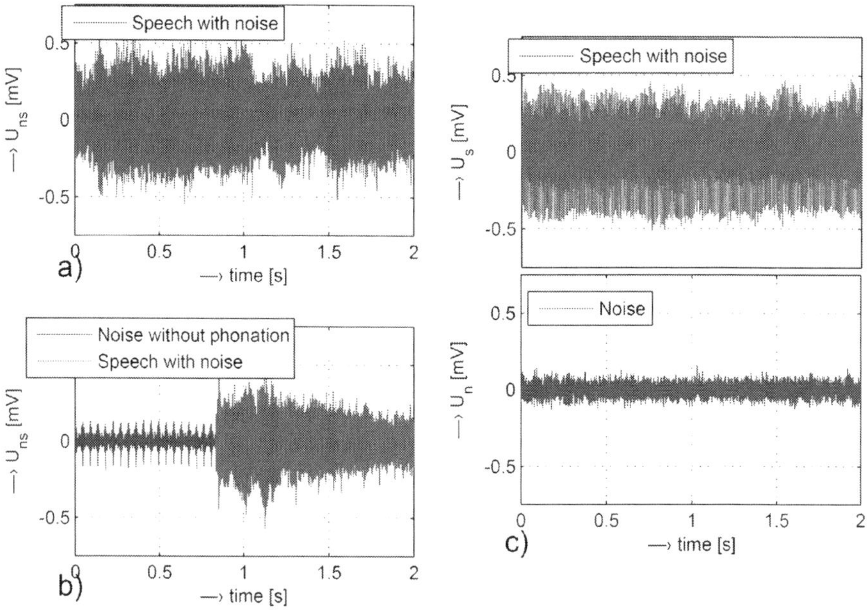

Figure 11. Examples of typical speech and noise waveforms for three analysed noise reduction methods: one-channel recording of the speech with noise (a), one-channel consecutive recording of an acoustic noise (without phonation) and the superimposed speech signal with noise (b) and the speech with noise and the noise signals recorded by two microphones in two separate channels during phonation of a vowel "a" by a male person (c), the used scan MR sequence 3D SSF.

Two corpora of the recorded signals were created by the described methods. The first one consists of the recorded noise signals only. These signals were used in two auxiliary experiments:

http://dx.doi.org/10.5772/64628

1. Analysis of noise supplementary spectral properties for different weights of the examined persons lying in the MRI device scanning area.

2. Analysis of the influence of the basic scan parameters (TR and TE times of the used MR sequence) on the spectral properties of the generated acoustic noise.

The noise signals brought about by one type of the MR scan sequence with different settings of TR = {10, 20, 30 and 40 ms} and then TE = {10, 14, 18 and 22 ms} were processed. From the recorded pure noise signals, the spectral envelopes and supplementary spectral properties were calculated and compared. The obtained values were next analysed statistically, and their histograms were calculated.

Person	**M1**$_{(JP)}$	**M2**$_{(TD)}$	**M3**$_{(LV)}$	**F1**$_{(AP)}$	**F2**$_{(BB)}$	**F3**$_{(ZS)}$
F0 $_{mean}$ **[Hz]**	133	127	98	228	177	207
Weight [kg]	78	75	80	53	50	55

Table 1. The mean *F*0 values and approximate weights of the persons participated in the experiment.

The second corpus consists of the speech signal collected in dependence on the tested noise reduction method: one or/and two channels—using only one or/and two pick-up microphones for signal recoding. They all originate from the recordings of six voluntary persons lying in the MRI scanning area (3 + 3 male/female) with different mean *F*0 values and different approximate weights—see **Table 1**. The main database of long Czech and Slovak vowels was consequently used for creation of the second one consisting of manually selected ROIs corresponding to the stationary parts of the vowels 'a', 'e', 'i', 'o' and 'u' (typically three to five parts with the mean duration of 3 s). Due to different *F*0 values of the persons participating in our experiment, different parameters of the window length w_L and the window overlapping w_O must be set—we use 24-ms frames for processing of the male voice speech signal and 20-ms frames for the female one. The parameter N_0 for limitation of the real cepstrum was chosen in correspondence with the period of the noise part of the signal L_N equal to 256 (when N_{FFT} = 1024) for both types of voices. The whole comparison process consists of three steps:

1. Visual comparison of the calculated spectrograms and periodograms using different frequency ranges: up to 2.5 kHz or in the whole bandwidth up to half the sampling frequency.

2. Numerical matching of the spectral distances as well as the signal RMS values.

3. Localization and determination of the maximum difference ΔP_{max} between spectral envelopes of the original noisy speech signal and the signal after noise suppression (in the low-frequency band) for final comparison of effectiveness of all three described noise reduction methods.

4. Discussion of results

From the performed auxiliary measurements of the acoustic noise SPL, directional pattern in the three distances follows that the maximum level of about 72 dB was achieved for the nearest location of the SPL meter (D_L = 45 cm), while the background noise SPL_0 originated from the temperature stabilizer reached approximately 52 dB (measured in the time instant when the no scan sequence is executed)—see the left graph in **Figure 12**. To prevent any interaction between the ferromagnetic pieces of the measuring devices (it holds mainly for the recording microphones) and the stationary magnetic field of the MRI equipment, we eliminated this close distance for future use in our experiments. On the other hand, differences between the measured SPLs in 60- and 75-cm distances were small; therefore, we finally chose the middle distance of D_L = 60 cm as the basic one for next measurements and recording experiments. Obtained results of the second auxiliary experiment are in the correspondence with our expectation that the maximum SPL should be achieved at the height h = 85 cm (in the middle between the upper and the lower gradient coils and the permanent magnets of the MRI device) where the noise from two components is superimposed, while for the SPL meter located at the heights of 75 and 95 cm, the noise distribution is practically the same and about 3 dB lower than at the middle height—compare the right directional pattern in the **Figure 12**.

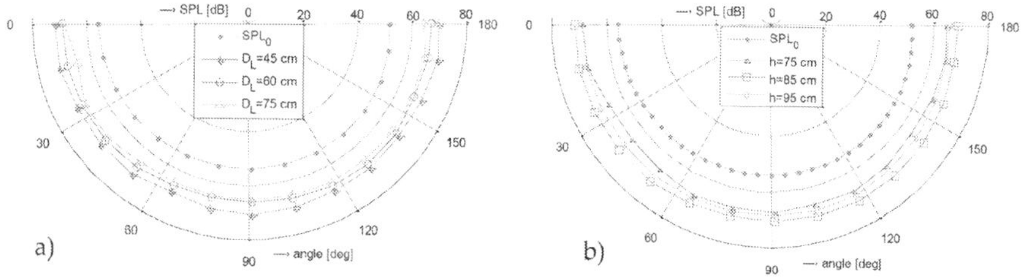

Figure 12. Measured directional patterns of the MRI gradient coils noise together with the background noise SPL_0: for three SPL meter distances at the height of 85 cm (a) and for three SPL meter heights at the distance of D_L = 60 cm (b).

Within the next auxiliary measurement, the influence of a weight of a lying person/water phantom in the scanning area of the MRI device on the vibration and subsequently generated acoustic noise SPL was analysed. The performed comparison of SPL measurement first of all shows that the maximum values are obtained at the position of 30° where the noise produced by the temperature stabilizer contributes for the most part of the resulting acoustic noise, and the minimum SPLs are achieved for the position of 15° with the maximum distance from the stabilizer. For all three measured directions, the noise SPL values measured with the examined person lying in the MRI scanning area were about 10 dB lower in comparison with the situation when the water phantom was used. When male and female testing persons lay on the bottom plastic holder of the permanent magnet and gradient coils, the obtained noise SPL values were roughly inversely proportional to their weights: the group of males with approximate weights of 78 kg and the group of females with the mean weight of 53 kg—as documented by the bar graph in **Figure 13**.

http://dx.doi.org/10.5772/64628

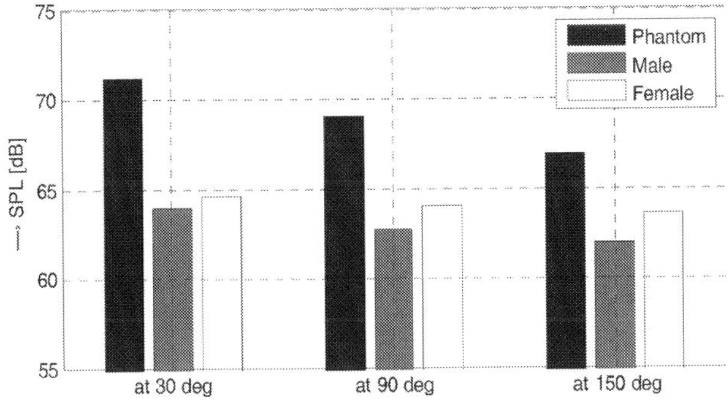

Figure 13. Comparison of the noise SPL values measured in the directions of 30, 90 and 150° using the water phantom, male and female persons lying in the scanning area of the MRI device without phonation, the SPL meter in the height of h = 85 cm, at the distance of D_L = 60 cm.

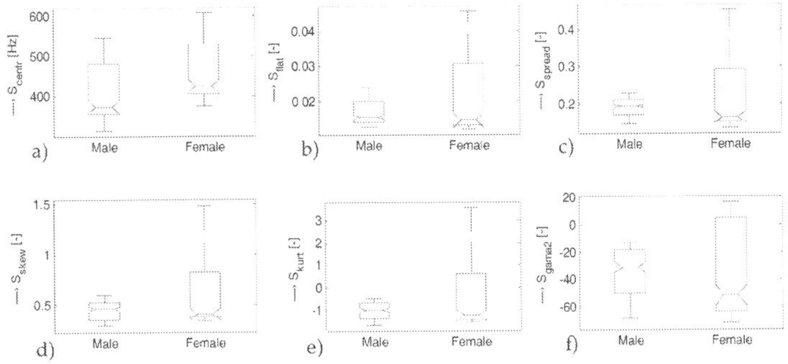

Figure 14. Box-plot of basic statistical properties of the basic and supplementary noise spectral parameters summarized for the groups of male/female persons lying in the scanning area of the MRI device: centroid (a), flatness (b), spread (c), skewness (d), kurtosis (e) and gama2 (f), signals recorded at the position of 150°.

Also another effect can be observed when the examined person lies in the scanning area of the open-air MRI and the holder of the lower gradient coils is loaded with his/her weight. The changed mass of the whole mechanical system affects its mechanical resonance modes, and the altered spectrum of the generated acoustic noise is apparent. To minimize the acoustic noise component caused by the temperature stabilizer, for further analysis of this phenomenon, the recording microphone was placed at 150° in the main experiment. The performed evaluation comprises comparison of the supplementary spectral properties for the weight pairs of male and female person groups—the box-plot of the basic statistical parameters (minimum, maximum, mean value and standard deviation) is shown in **Figure 14** and calculated histograms of the noise spectral properties is shown in **Figure 15**. The differences are most significant in the case of the spectral centroid feature with typical peaks of the histogram located around 375 Hz (for the male person group) and near 425 Hz for the female testing person group. In addition, the deviation of the obtained values is higher for the female person group than for the male one. It can be explained by greater differences among the weights within this group (compare the approximate values in **Table 1**).

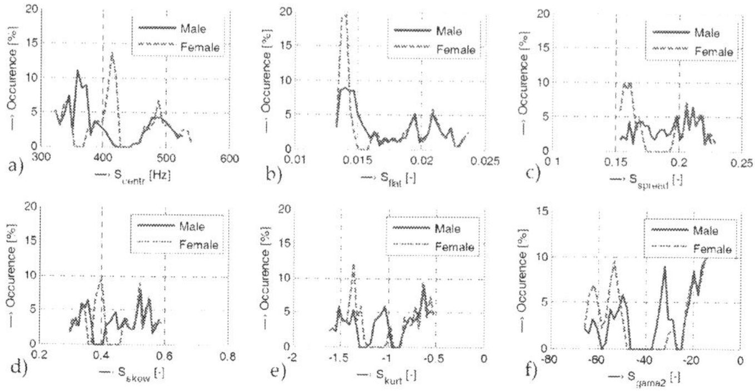

Figure 15. Histograms of the noise spectral properties summarized for the groups of male/female persons lying in the scanning area of the MRI device in correspondence with **Figure 14**: centroid (a), flatness (b), spread (c), skewness (d), kurtosis (e) and gama2 (f).

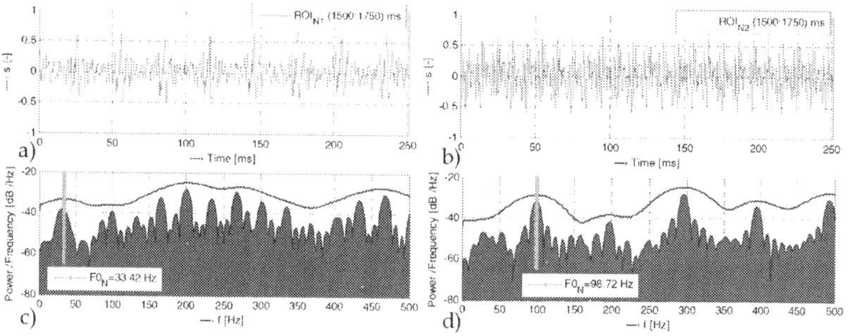

Figure 16. Detailed analysis of the influence of different TR parameter settings on changes in the basic spectral properties of generated noise signals: examples of vibration signals in the selected ROI in 250-ms time interval for TR = 33 ms (a) and TR = 10 ms (b), corresponding spectral densities including the spectral envelopes depicted for the frequency range up to 500 Hz together with the determined basic $F0_N$ frequencies (c and d), the used scan sequence of 3D SSF type with setting TR = 10 ms.

The aim of the last auxiliary experiment was to analyse how the setting of the basic scan parameters of the used MR sequence (TR and TE times) impacts the spectral properties of the generated acoustic noise. The performed detailed analysis shows that different settings of TR and TE parameters are manifested by changes of the noise spectral properties as follows:

- The repetition time has much influence on the basic $F0_N$ frequency of the generated periodical noise as documented by the graphs of the spectral density together with the smoothed spectral envelopes as well as by the results of the statistical analysis of the $F0_N$ frequencies for different TR time settings (see **Figures 16** and **17**).

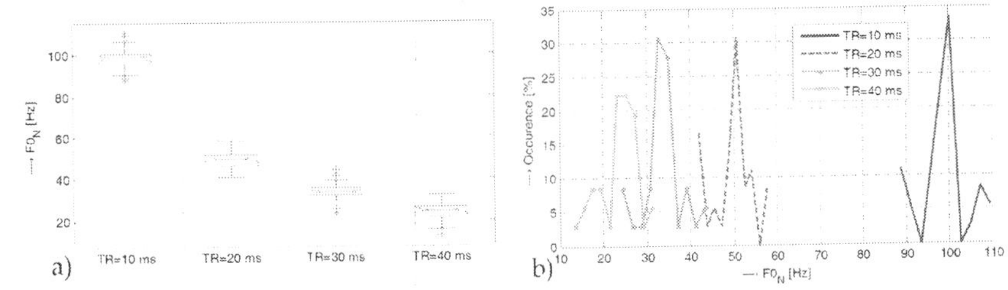

Figure 17. Summary comparison of influence of different TR parameter settings on the basic spectral properties of generated noise signals: box-plot of basic statistical properties of the determined basic $F0_N$ frequencies for TR = {10, 20, 30 and 40 ms} (a) and corresponding histograms (b), the processed scan sequence of 3D SSF type with TE = 10 ms.

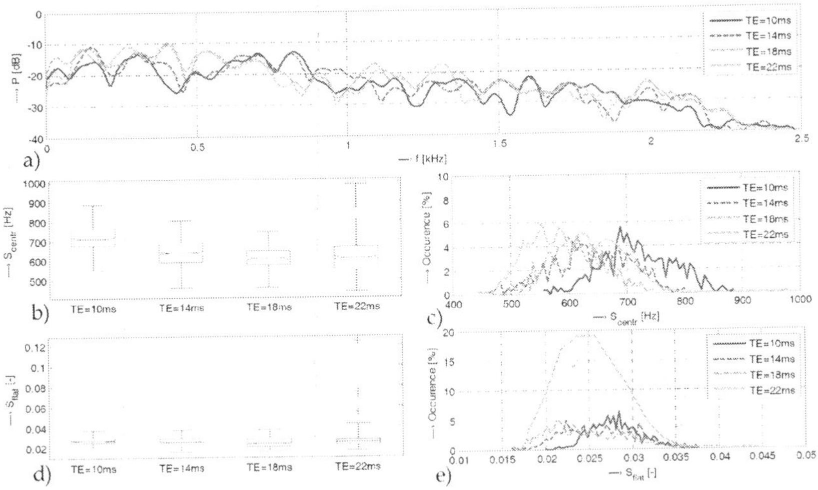

Figure 18. Influence of different TE parameter settings on the spectral properties of the generated noise: spectral envelopes depicted for the low-frequency band up to 2.5 kHz (a), box-plot of statistical parameters of S_{centr} (b), histograms of distribution of S_{centr} (c), box-plot of statistical parameters of S_{flat} (d) and histograms of distribution of S_{flat} (e), the used scan sequence of 3D SSF type with setting of TR = 30 ms and TE = {10, 14, 18 and 22 ms}.

- Echo time settings have effect on higher frequencies of the noise signal, as it is well documented by its supplementary spectral properties—see the box-plot of the basic statistical parameters of the centroid and the flatness together with their histograms in **Figure 18**.

The performed main comparison experiment confirms usability of all three applied noise reduction methods based on cepstral limitation and clipping, or using one-channel/two-channel spectral subtraction. The significant differences between the noisy and the cleaned speech signals were observed for all processed samples as documented by the graphs in **Figures 19–21**, and by the resulting mean, spectral distances were calculated between spectral envelopes of vowels recorded in MRI noisy environment and cleaned by both noise reduction methods summarized in **Table 2**, as well as the calculated mean spectral differences ΔP at the basic noise frequency $F0_N$ for all analysed stationary parts of vowels listed in **Table 3** (for both speaker genders). Comparison of calculated spectral distances between the original noisy and the cleaned speech signals shows that spectral changes are higher when using the second noise reduction method. However, in practical realization, it is very difficult to fulfil the basic condition—to find a suitable arrangement of the Mic. 2 for the pure noise recoding. It means that there are some recording imperfections resulting in a crosstalk of the speech signal into the Mic. 2 intended only for noise recording. Next disadvantage of the spectral envelope subtraction method consists in the fact that this approach brings also undesirable effects (e.g. aliasing), and therefore, some frequency filtering must be subsequently carried out. This shortcoming can be solved also in the cepstral domain—by greater limitation of the real cepstrum expansion, but it would be contradictory to the requested higher quality of the finally reconstructed speech signal.

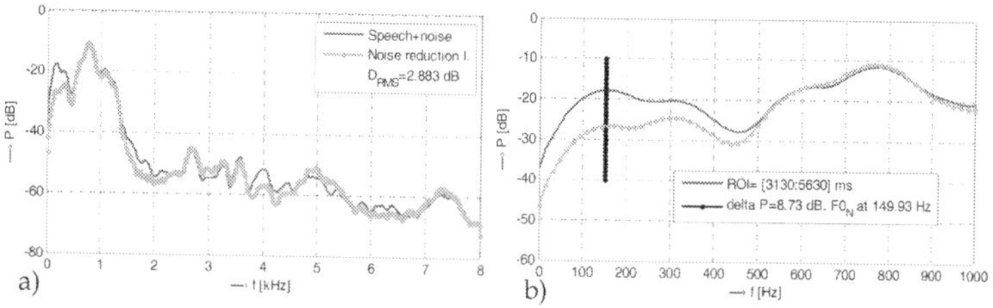

Figure 19. Comparison of spectral envelopes for noise reduction method I: summary view in the full frequency range up to $f_s/2$ (a), detailed part for the frequency range up to 1 kHz with the determined spectral difference ΔP at the frequency $F0_N$ (b), processed speech signal of the long vowel "o" (female voice) in the selected ROI of 2.5-s time interval, the used scan sequence of 3D SSF type, TR = 8 ms, f_s = 16 kHz.

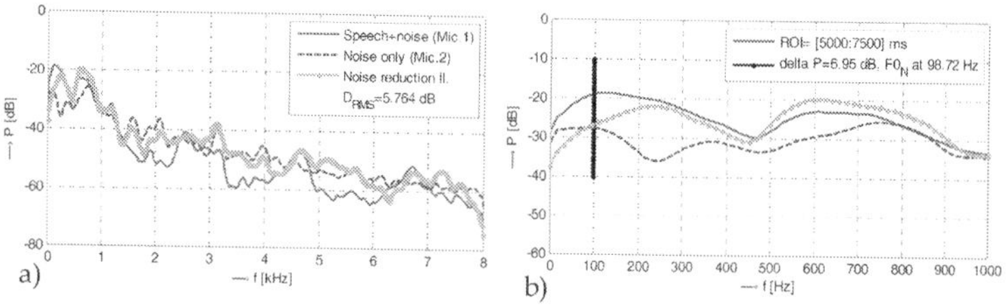

Figure 20. Comparison of spectral envelopes for noise reduction method II: summary view in the full frequency range up to $f_s/2$ (a), detailed part for the frequency range up to 1 kHz with the determined spectral difference ΔP at the frequency $F0_N$ (b), processed speech signal of the long vowel "a" (male voice) in the selected ROI of 2.5-s time interval, the used scan sequence of 3D SSF type, TR = 10 ms, f_s = 16 kHz.

http://dx.doi.org/10.5772/64628

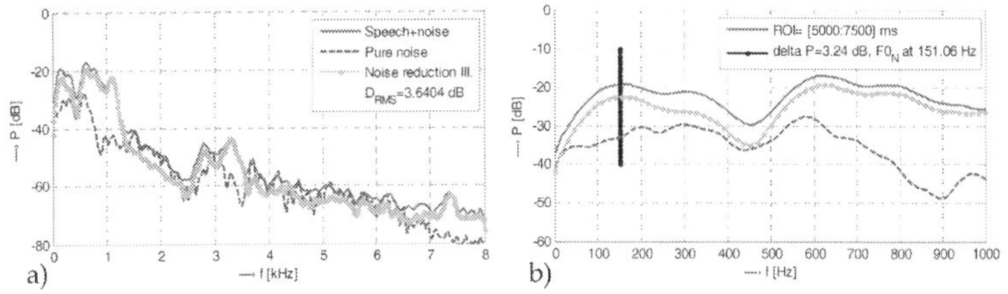

Figure 21. Comparison of spectral envelopes for noise reduction method III: summary view in the full frequency range up to $f_s/2$ (a), detailed part for the frequency range up to 1 kHz with the determined spectral difference ΔP at the frequency $F0_N$ (b), processed speech signal of the long vowel "o" (female voice) in the selected ROI of 2.5-s time interval, the used scan sequence of 3D SSF type, TR = 8 ms, f_s = 16 kHz.

Stationary part of vowel	Male voice			Female voice		
	Method I	Method II	Method III	Method I	Method II	Method III
a: D_{RMS} [dB]	7.32	5.58	4.21	5.45	5.33	3.87
e: D_{RMS} [dB]	6.49	5.24	3.97	4.86	5.01	3.82
i: D_{RMS} [dB]	7.36	5.63	4.14	5.27	5.14	3.10
o: D_{RMS} [dB]	5.66	4.98	4.02	6.11	4.86	3.19
u: D_{RMS} [dB]	4.73	4.89	3.81	5.46	4.78	4.42

Table 2. Comparison of the mean values of spectral distances of the analysed vowels (D_{RMS} are calculated between spectral envelopes of the original noisy and the cleaned speech signals) for male and female voices.

Stationary part of vowel	Male voice			Female voice		
	Method I	Method II	Method III	Method I	Method II	Method III
a: ΔP [dB]	6.38	7.22	3.26	4.89	6.72	2.84
e: ΔP [dB]	5.79	6.59	2.95	5.12	5.84	2.86
i: ΔP [dB]	6.11	7.23	3.45	5.64	6.18	3.28
o: ΔP [dB]	5.07	5.86	3.08	4.85	5.68	3.11
u: ΔP [dB]	5.58	4.97	2.86	5.34	5.02	2.85

Table 3. Results of mean spectral differences ΔP at the basic noise frequency $F0_N$ (determined between the spectral envelopes of the original noisy and the cleaned speech signals) for male and female voices.

Noise suppression based on cepstral peaks clipping can also cause suppression in the spectrum at the fundamental speech frequency $F0$ or at the first formant frequency $F1$ if they are close to the clipped noise frequency $F0_N$. Generally, it holds that the values of these frequencies $F0$ and $F1$ depend merely on the speaker vocal tract characteristics and they may not be affected by noise suppression in the speech signal. The third noise suppression method based on the noise estimation techniques using statistical evaluation of the spectral properties has a principal handicap—this approach is not able to track real variations in the noise and its application in noise suppression results in an artificial residual fluctuating noise and a distorted speech.

5. Conclusions

The achieved results will serve to create databases of initial parameters (such as the filter bank for noise signal pre-processing) used to design a filter for noise suppression in the speech signal recorded simultaneously with 3D human vocal tract scanning. It will be useful in experimental practice when it often occurs that the basic parameter setting of the used scanning sequence as well as the other scanning parameters must be changed depending on the person that is currently examined. The advantage of the system is in the design method of the device noise elimination by specially developed software which enables to measure the acoustic characteristics of the subject during phonation in the MRI device with good accuracy. Finally, these three described algorithms have much a wider range of applications than indicated by the particular experimental arrangement in this paper.

For better knowledge of the acoustic noise conditions in the scanning area and in the vicinity of the MRI device, it is necessary to carry out additional measurement and experiments. The results obtained in this way will help to describe the process of the gradient coil electric excitation, the subsequent mechanical vibration and the resulting acoustic noise generation in the MRI device scanning area and its neighbourhood. There is also need for the knowledge about the contribution of the upper gradient coil (and its plastic holder) to the resulting acoustic noise. For this reason, the parallel measurement of the vibration signal on the surface of both plastic holders as well as the analysis of reflective properties of the metal shielding cage and its influence on the progression of the acoustic wave should be performed.

Acknowledgements

The work has been done in the framework of the COST Action IC 1206 and has been supported by the Grant Agency of the Slovak Academy of Sciences VEGA 2/0013/14.

Author details

Jiří Přibil[1*], Anna Přibilová[2] and Ivan Frollo[1]

*Address all correspondence to: umerprib@savba.sk

1 Institute of Measurement Science, SAS, Bratislava, Slovakia

2 Faculty of Electrical Engineering and Information Technology, SUT, Bratislava, Slovakia

References

[1] Vampola T, Horáček J, Laukkanen AM, Švec JG. Human vocal tract resonances and the corresponding mode shapes investigated by three-dimensional finite-element modelling based on CT measurement. Logopedics Phoniatrics. 2015; 40:14–23.

[2] Zhu Y, Kim YC, Proctor MI, Narayanan SS, Navak KS. Dynamic 3-D visualization of vocal tract shape during speech. IEEE Transactions on Medical Imaging. 2013; 32:838–848.

[3] Burdumy M. et al. Acceleration of MRI of the vocal tract provides additional insight into articulator modifications. Journal of Magnetic Resonance Imaging. 2015; 42:925–935.

[4] Moelker A, Wielopolski PA, Pattynama MT. Relationship between magnetic field strength and magnetic-resonance-related acoustic noise levels. Magnetic Resonance Materials in Physics Biology and Medicine. 2003; 16:52–55.

[5] Kannan G, Milani AA, Panahi IMS, Briggs RW. An efficient feedback active noise control algorithm based on reduced-order linear predictive modeling of fMRI acoustic noise. IEEE Transactions on Biomedical Engineering. 2011; 53:3303–3309.

[6] Shou X. et al. The suppression of selected acoustic frequencies in MRI. Applied Acoustics. 2010; 71:191–200.

[7] Wu Z, Kim YC, Khoo MCK, Nayak KS. Evaluation of an independent linear model for acoustic noise on a conventional MRI scanner and implications for acoustic noise reduction. Magnetic Resonance in Medicine. 2014; 71:1613–1620.

[8] Přibil J, Horáček J, Horák P. Two methods of mechanical noise reduction of recorded speech during phonation in an MRI device. Measurement Science Review. 2011; 11:92–98.

[9] Montazeri V, Pathak N, Panahi I. Two-channel multi-stage speech enhancement for noisy fMRI environment. Canadian Journal of Electrical and Computer Engineering. 2013; 36:60–67.

[10] Sun G, Li M, Rudd BW, Lim TC, Osterhage J, Fugate EM, Lee JH. Adaptive speech enhancement using directional microphone in a 4-T scanner. Magnetic Resonance Materials in Physics, Biology and Medicine. 2015; 28:473–484.

[11] Wang J, Liu H, Zheng C, Li X. Spectral subtraction based on two-stage spectral estimation and modified cepstrum thresholding. Applied Acoustics. 2013; 74:450–458.

[12] Vích R, Přibil J, Smékal Z. New cepstral zero-pole vocal tract models for TTS synthesis. In: IEEE Region 8 EUROCON'2001; Bratislava, Slovakia, 2001. p. 458–462.

[13] Aalto D. et al. Large scale data acquisition of simultaneous MRI and speech. Applied Acoustics. 2014; 83:64–75.

[14] Narayanan S. et al. Real-time magnetic resonance imaging and electromagnetic articulography database for speech production research. Journal Acoustical Society of America. 2014; 136:1307–1311.

[15] Přibil J, Přibilová A, Frollo I. Analysis of spectral properties of acoustic noise produced during magnetic resonance imaging. Applied Acoustics. 2012; 3:687–697. doi:10.1016/j.apacoust.2012.01.007

[16] Přibil J, Přibilová A, Frollo I. Influence of Person Weight on Spectral Properties of Acoustic Noise in the Open-air MRI. In: Holub J, editor. XXI IMEKO World Congress "Measurement in Research and Industry" – Full Papers; August 30–September 4, 2015; Prague, Czech Republic. Issue 1, 2015. 4 p.

[17] Esaote E-scan Opera. Image Quality and Sequences Manual. 830023522 Rev. A. Esaote S.p.A., Genoa, April 2008.

Chapter 10

Noise Reduction and Control in Hospital Environment: Design of the NeoNoise Project

Carlos Carvalhais, Manuela V. Silva, Ana Xavier and Joana Santos

Additional information is available at the end of the chapter

http://dx.doi.org/10.5772/64629

Abstract

The "NeoNoise Project: Integrated Approach to Minimize Sound Pressure Levels in Neonatal Intensive Care Units" is being conducted by the Research Group on Occupational and Environmental Health of the Research Center on Health and Environment in neonatal intensive care units, since exposure to sound pressure levels in these spaces has been recognized as a factor that influences the quality and well-being of the occupants (workers and others), as well as the recovery of premature infants who are hospitalized. This work reports the rationale and the design of the NeoNoise project as well as the methods used for data collection. A brief review on the results published and available for the scientific community is also made. In general, NeoNoise project intends to make an integration of all relevant factors, with the intention of presenting a guiding document to change the working practices and occupant's behaviors. So far, this study provided data on sound pressure levels by objective and subjective approaches, as well as information about the exposure factors and sensitivity of the occupants to noise.

Keywords: premature infants, study protocol, neonatology, noise effects, noise perception

1. Introduction

Noise is an environmental stressor that is known to have physiological and psychological effects. The body responds to noise in the same way it responds to stress and overtime has potential to impair health. In general, vulnerable groups are underrepresented in study populations. Although anyone might be adversely affected by noise exposure (environmental or occupational exposure), groups that are particularly vulnerable include neonates, infants, children, those with mental or physical illnesses, and the elderly. In hospital environment, excessive noise is not only annoying, but can also interfere with the proper performance of health care. Evidence shows that hospital noise levels often exceed those recommended by World Health Organization (WHO) [1] and other agencies. In hospitals, patient exposure has been studied more frequently over the years, than professional exposure. However, the particular case of neonatal intensive care units (NICUs) poses a new challenge, due to the "type" of patients involved—ill and/or premature infants. Newborn infants who need intensive medical attention are often admitted into an NICU. These units combine advanced technology and trained healthcare professionals to provide specialized care for ill and/or premature newborns. NICUs may also have intermediate or continuing care areas for babies who are not as sick but do need specialized health care. Noise production in NICU rooms and inside incubators is usually due to alarms produced by life support devices, flow of medical gas, communication among professionals/visitors and during activities of nursing care [2–4]. **Table 1** shows the main causes of noise in NICU.

Source of noise	
Items falling onto the floor	Up to 92 dB(A)
Equipment movement (e.g., bed)	90 dB(A)
Connection of gas supply	88 dB(A)
Door closure	85 dB(A)
Pager	84 dB(A)
Talking	75–85 dB(A)
Ventilator alarm	70–85 dB(A)
Nebulizer	80 dB(A)
Telephone	70–80 dB(A)
Television	79 dB(A)
Oximeter	60–80 dB(A)
Monitor alarm	79 dB(A)
Ventilator	60–78 dB(A)
IV infusion alarm	65–77 dB(A)
Endotracheal aspiration unit	50–75 dB(A)

Table 1. Equipment and behavioral causes of noise in intensive care units [5].

Health professionals are aware about this issue and identified noise as an agent with a negative impact on work performance [6–8]. In fact, it is known that the hospital environment has many occupational health risks due to the variety of clinical and nonclinical tasks performed by healthcare workers. The exposures to psychosocial, chemical, physical, mechanical, and biological hazards are common in hospital units and predispose healthcare workers to different types of accidents [9]. However, the work performed in NICU can be particularly psychologically demanding which combined with noise exposure within the NICU can increase the risk of work accidents occurrence, with negative consequences for staff and also for patients. In fact, noise may induce extraauditory effects in professionals including burnout, stress, and fatigue [10]. There is some association between noise and some health outcomes such increases in blood pressure, heart rate, hypertension, and other cardiovascular diseases. Noise exposure can also stimulate the release of epinephrine (adrenaline), increase pain, and alter quality of sleep [11, 12]. Even in newborns these effects are being implicated and associated with noise [13]. Although, it is important to underline that the levels of noise exposures associated with these health effects range widely [14].

2. Rationale and aim of the project

A literature review conducted by Konkani and Oakley [15] showed that several authors studied and characterized acoustic environment of intensive care units in hospitals. Studies measuring noise amplitude in dB or frequency analysis or through an approach combining noise measurements and patient or staff questionnaire surveys or interviews are quite usual in this domain. Dube et al. [16] surveyed patients to identify the noisiest time of the day, and were also asked to list the noises that they felt were annoying. Connor and Ortiz [17] conducted a survey where patients rated the noise level before and after a staff education program. However, to our knowledge in Portugal, until 2010 only one study was performed in intensive care units, namely in NICU. Nicolau et al. [13] characterized noise levels in six NICUs of Lisbon region, revealing that noise levels were above the recommended by international guidelines. They emphasized the need to train healthcare staff and include actively health professionals in noise reduction strategies. Due to the lack of data in Portugal, including lack of studies measuring the effectiveness of noise reduction strategies, the Research Group on Occupational and Environmental Health of the Research Center on Health and Environment (SOA/CISA), designed the "NeoNoise Project: Integrated Approach to Minimize Sound Pressure Levels in Neonatal Intensive Care Units." NeoNoise project intends to be a contribution to understand the role of educational, environmental, and infrastructural factors on noise reduction and health promotion in neonatal intensive care units considering two major risk groups: premature infants and staff. The specific goals of the project are (1) to characterize sound pressure levels in different locations of NICU; (2) to determine the influence of these levels on health and well-being of premature infants and health professionals (3) to identify staff perceptions regarding working conditions, comfort, and main noise sources in NICU; (4) to develop and implement a quiet time protocol in NICU; (5) to study the relationship between the previous factors; (6) to create a good practices guide for these environments, in order to control noise production and improve well-being, comfort and satisfaction levels of professionals and patients; and finally (7) to suggest recommendations for health authorities, the scientific community and general public.

The main goal of this work is to present and discuss the study design and protocol of the NeoNoise project, by reviewing its rationale and outlining methods that might be implemented by other researchers in this field.

3. Materials and methods

This project started in 2011 and is being conducted in three NICUs located in hospitals of the North of Portugal involving some tasks/activities that were or will be performed simultaneously in order to complete the study. NeoNoise was designed to be carried out in two different phases. The study protocol and the concluded and ongoing substages/studies (underlined in red) are presented in **Figure 1**.

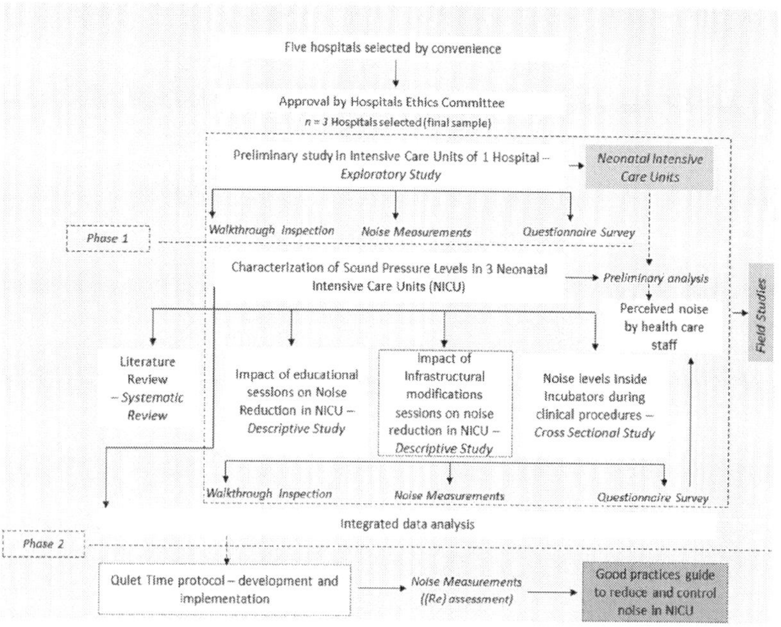

Figure 1. NeoNoise protocol.

3.1. Selection of the NICU and ethical issues

The north region of Portugal has six public hospitals with differentiated perinatal support. Five hospitals were contacted, and authorization to perform the study was given by three hospitals, after favorable statement by their Ethics Committee and approval by their respective administration boards. The study was carried out according to the Helsinki Declaration. **Figure 2** shows the location of these three hospitals.

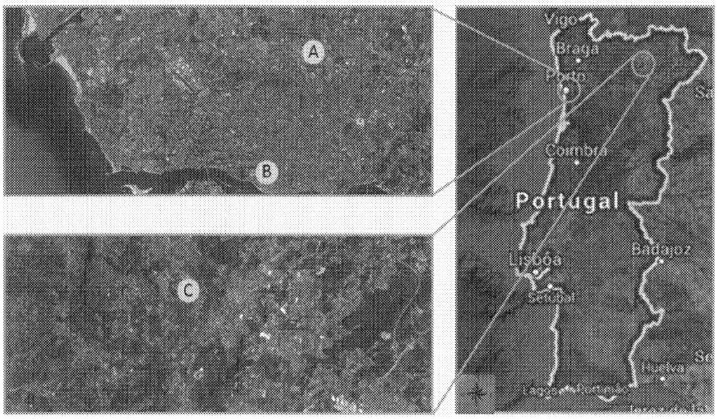

Figure 2. Spatial distribution of the three hospitals involved in the study (A and B in Porto and C in Vila Real, Portugal).

3.2. Field investigations in the NICU

As shown in **Figure 1**, field investigations were transversal to the most of the substages of the project. They included walkthrough inspections and assessment of sound pressure levels in the different spaces of the selected NICU. Additionally, healthcare staff answered a self-reporting questionnaire. In order to perform the ongoing tasks, some general considerations about methodological procedures are made below.

3.2.1. Walkthrough inspections

Walkthrough inspections were made by two trained researchers, in order to characterize the built environment and indoor spaces of the three NICUs under study. A checklist for this purpose was used. It should be noted, that since in Portugal there is no legislation related to NICU design for public institutions, the checklist was based on legal requirements applicable to private healthcare units, which have specific criteria for the design, conception, and equipment that should exist in NICU. Detailed information regarding the building environment such as traffic and rural/urban surroundings and other external noise sources, construction characteristics, among others, was gathered. Identification of all relevant information such as area, finishing materials, and conditions concerning floor, walls, ceiling, windows, and ground as well as equipment installed and healthcare routines was made. Partial information about the characteristics of NICU is presented in **Table 2**. Detailed information is given in [3, 18, 19].

NICU	General characterization
A	14 incubators, 5 nurseries, 2 workstations, 4 sinks, 1 isolation room, 1 waste storage room, 1 storage room, 1 meeting room.
B	6 (or 7)* incubators, 3 nurseries, 1 workstations, 5 sinks, 2 isolation room, 1 storage room, 1 meeting room.
C	11 incubators, 8 nurseries, 2 workstations, 4 sinks, 1 milk preparation room, 1 WC, 1 storage room, 1 meeting room.

Note. *When necessary, one more incubator can be installed.

Table 2. General characteristics of the three NICUs.

3.2.2. Noise measurements

The measurements were mostly carried out continuously over 24 hours, during seven days in each measurement place (work station, traffic zone, inside incubator). Inside the incubator, short measurements (5–10 min.) were also made. The measurement protocol was based on the orientations of previous studies [20]. In this sense, a preliminary survey was performed in order to identify noise sources. Measurements were performed using a sound level meter class 1 (01 dB®, model Solo-Premium). The measurements of peak sound pressure level (Lp, Cpeak) were made using the C filter and the A-weighted equivalent sound pressure level (LAeq) were obtained using the A filter, which is a frequency weighting filter that simulates human hearing. "C-weighting" curve was used, providing a flat frequency response with slight attenuation for high and low frequencies. It is usual to measure the peak noise levels in hospitals environment in order to define improvements to the acoustical environment [21]. Slow response time averaging (1 s) was also used because it is the most appropriate response for the majority of the applications in hospitals and provides stable readings [22]. To ensure accurate measurement, recording was preceded by calibration of the sound level meter [23] with an acoustic calibrator class 1 (RION®, model NC-74). In the analysis and interpretation of results reference values given by WHO [1], were used. **Table 3** shows reference levels for hospitals, given by WHO and other organizations. After the field measurements, the data were transferred and processed in the dBTRAIT software, version 5.4.

Organization	Recommended values
United States Environmental Protection Agency [24]	45 dBA daytime 35 dBA night
World Health Organization [1]	For areas where patients are treated or observed—35 dB LAeq For wardrooms in hospitals—30 dBA LAeq with a corresponding LAmax (maximum A-weighted sound pressure level) of 40 dBA
Committee on Environmental Health—American Academy of Pediatrics [25]	45 dBA

Table 3. Recommended noise levels in hospitals.

3.2.3. *Questionnaire survey*

The analysis of staff noise perception in their workplaces involved the application of a questionnaire, in order to characterize working conditions, comfort, and the main noise sources. The questionnaire developed and tested in previous studies of this project, was divided into three main sections containing a total of 11 questions: (1) demographic information (sex, age, profession, years of work in NICU, shift); (2) judgment of personal acceptability of noise and comfort; and (3) judgment of the noisiest shift and main sources of noise in the NICU. There were no contacts between the researchers and the participants. The questionnaire was delivered and received by a nurse, responsible for the NICU. The questionnaire fulfillment was completely anonymous and confidential. This questionnaire was (and will be) used in different studies of the project. Information regarding noise perception by professionals during the completed stages is given by [18]. Other results and respective data analysis regarding questionnaire survey are being considered for another publication.

3.3. Literature review

This task consisted in a short systematic review, conducted in selected databases and based on PRISMA statement [26], to summarize studies characterizing noise levels in hospital NICUs, in the last 15 years (since the year 2000), to gather more relevant and recent information. Some of the keywords used were NICU, noise and hospital, noise, among others. It was an important study, in order to determine gaps in knowledge and to define the purpose and concept of the NeoNoise project, more accurately.

3.4. Behavioral and structural modifications in NICU

The activities regarding behavioral changes were already performed. In this phase of the project measurements were made before and after a training program (TP) in one NICU. TP was conducted by three researchers. The TP was performed through a lecture of approximately 60 min and conducted by the investigators. To ensure that all the staff of the NICU under study such as physicians, nursing staff, and auxiliary staff attended the lecture (n = 79), 14 training sessions were given [2, 6]. The lecture included (1) general concepts of noise; (2) the results of the sound pressure levels obtained in the first phase and the comparison of these to the recommended values suggested by WHO and other regulatory agencies; (3) the negative impact of noise on health, both for neonates and professionals; and (4) some actions that needed to be implemented to ensure noise reduction were undertaken [6]. Regarding these actions, health professionals had a significant role in the development of an action plan to address specific noise issues. Detailed information is given in [6].

The tasks regarding the effectiveness of environmental or infrastructural modifications will be conducted in one NICU. As referred before, this field investigation will involve a walkthrough inspection by two trained researchers using a checklist and measurements for the assessment of the sound pressure levels. Some infrastructural modifications are being performed in the selected NICU for this study (B). Noise measurements were already made before and will be carried out after these modifications.

3.5. Quiet time protocol (QTP)

Based on acquired knowledge obtained in all the studies developed within the scope of NeoNoise project, it will be developed a quiet time protocol involving not only frequent and ongoing training sessions of healthcare staff, but also other good practices to control noise production and guarantee a quiet environment. Quiet times are designated hours where activity and conversation is minimized to allow patients to rest. Some authors referred the most effective model is to have a period in the afternoon and a period during the night, when quiet hours are observed. However the structure of the quiet times must to be defined taking into account shift changes, among other specific activities of the NICU. Quiet hours could be observed in many ways (when possible), such: conduct conversations in workstations and other areas in a hushed manner; encourage visitors to participate and also to take breaks to let patients rest; restrict phone conversations to designated areas of the NICU; minimize or eliminate clinical interventions (e.g., blood draws, etc.) during these hours, etc. The effectiveness of QTP will be tested in three NICUs, through noise measurements and questionnaire survey.

3.6. Good practices guide

Based on previous phases of NeoNoise project and taking into account the reality of the Portuguese healthcare services, a manual will be developed and published. This will help health professionals in the adoption of efficient strategies to reduce the production of noise not only in NICU but also, in other intensive care units.

3.7. Data management and analysis

Data gathered during the project is being managed and analyzed through IBM SPSS™ (Statistical Package for the Social Sciences) 20th version and MS Excel® 2013 software's. Data obtained by measurements were transferred and processed in the dBTRAIT software version 5.4 and exported to MS Excel® 2013 for further analysis. Databases were developed specifically for the study by the research team in order to record the large amount of data. Data input was the responsibility of two researchers of the project. An exploratory analysis of the variables of interest was carried out using classic descriptive statistics to calculate frequencies, means, medians, and associated dispersion measures with analysis of LAeq and Lp, Cpeak values. Normality, parametric and nonparametric tests of hypothesis were also used as appropriate. All tests considered a 95% confidence interval.

3.8. Dissemination

As previously mentioned, some studies within the scope of NeoNoise project were published or submitted for publication in international peer reviewed journals and presented at international scientific conferences. The results were communicated to the NICU responsibles to better understand noise production and its sources and to contribute for the development of preventive measures, through technical reports and short information sessions. Additionally, a final conference/seminar will be organized to disseminate results to the general public.

4. Results and discussion

NeoNoise is the first Portuguese study addressing the effect of noise on premature infants and healthcare staff through objective measurements of sound pressure levels and subjective analysis by questionnaire surveys, and testing the effectiveness of different noise reduction strategies in the NICU. Data collection was carried out successfully (except for the stages that are not completed yet). Data analysis is still ongoing, but preliminary results were already presented at scientific meetings and published or accepted for publication. Formal recommendations to national authorities and public education materials will be made available in written documents.

In the exploratory study shown in **Figure 1**, Santos and Miguel [19] combined objective measurements of noise and a questionnaire survey in order to characterize noise levels in eight intensive care units (ICU) of a hospital, located in Porto, Portugal. The study also involved the application of the Ergonomic Workplace Analysis (EWA) methodology adapted by Miguel et al. [27] for the determination of risk level and intervention prioritization. The values of LAeq dBA obtained in the ICUs ranged from 50.0 to 65.0 dBA in the center of the units and between 57.8 and 67.1 dBA at the bedside of the patients. These values are above those recommended by WHO. Similar results were obtained by several authors in the same type of units [4, 8, 28, 29]. It is important to note that during the measurements, different operational equipment, including alarms, monitors, ventilators, infusion pumps, and nebulizers were operating. The conversation between the health professionals team at ICU was also identified as a possible source of noise that interfered the results. Comparing the results, it was found that the morning LAeq dBA values were higher than the afternoon ones, which may be related to the fact that during this period, medical examinations and hygiene of patients were more frequent. All ICUs had noise levels above the recommended and NICU was considered for further studies due to the patients involved: premature infants, who are not able to complain about noise. In fact, in Portugal there has been a considerable increase in preterm births, which in 2004 increased from 6.7 to 8.8% in 2009 [30]. Thus, it is essential to promote a quiet environment to reduce the impact of noise levels on health and well-being of premature infants and health professionals. In this sense, Santos et al. [3] documented some preliminary results on noise levels and responses given by healthcare staff of a NICU. It was found that during the week, the mean values of LAeq dBA obtained in the evaluated rooms ranged from 48.3 to 82.5 dBA. The results demonstrated that Monday LAeq dBA values were higher than the others days of the week, ranging between 52.0 and 86.0 dBA. Furthermore, sound pressure levels were significantly higher on weekdays than on weekend days ($p < 0.05$). In general, mean values of LAeq were lower in night shift; such was already reported for other authors [31, 32]. Night period is characterized by fewer visitors and health professionals and low lighting, which might reduce conversation. Significant differences have been found between the morning and night shift ($p < 0.05$) and between the afternoon and night shift ($p < 0.05$). On the other side, no significant differences has been found between the morning and afternoon shift ($p = 0.369$). Questionnaire survey showed that patient care activities and conversation between staff and visitors were identified as an important source of noise. This study concluded that noise levels were above the recommended and that routine activity of healthcare professionals has been identified as a potential source of noise. It was emphasized that training the staff in order to implement quiet work behaviors is essential, but changing physical elements of a space can result in great noise reduction. Following those conclusions, Carvalhais et al. [6] conducted a pilot study regarding the effectiveness of a training program on noise reduction in an NICU. The results showed that after six months of TP implementation, there was no significant noise reduction in the NICU rooms and inside the incubator. The "Work Station" of Room A had a decrease on LAeq and Lp, Cpeak values, 71.7–58.8 dBA and 143.3–102.8 dBC, respectively. However, in the "Traffic Zone" of Room B, the noise level increase almost 6 dB after the TP, probably attributed to the presence of visitors and other staff (from ancillary departments that did not participated in the TP) and might be the source of this rise. The LAeq values obtained in the "Work Station" and "Traffic Zone" before and after the implementation of TP exceed the recommended values given by WHO for day and night periods, indicating

	N (%)	**Mean (min-max)**
N	95 (100)	
Sex		–
Male	9 (9.5)	
Female	86 (90.5)	
Age in years		40.4
18–39	45 (47.4)	(24–61)
40–59	(48.4)	
>60	2 (2.1)	
missings	2 (2.1)	
Professional group		–
Operational assistants	24 (25.3)	
Nurses	52 (54.7)	
Physicians	17 (17.9)	
missings	2 (2.1)	
Years at NICU		10.1
<5	34 (35,8)	(0.5–35)
5–20	47 (49.5)	
>20	9 (9.5)	
missings	5 (5.3)	
Shift		---
Morning	53 (55.8)	
Afternoon	24 (25.3)	
Night	18 (18.9)	

Table 4. Demographic characteristics of the sample (n = 95) [18].

more attention needs to be taken. A spectral analysis was also made. In this study healthcare professionals (n = 79) were asked to identify the main sources of noise. Visitors, equipment, healthcare procedures, and conversation among others, were generally the most referred sources.

The workers perception in those environments is very important in the definition, development, and implementation of an intervention to reduce noise levels and to ensure that changes take place. In this sense, a questionnaire survey was performed in order to characterize health staff perceptions regarding noise in NICU [18]. A total of 95 professionals from three NICUs participated in this study. **Table 4** shows the characteristics of the sample.

The majority of the respondents (55.8%) found "equipment" (including telephones and the signals and sounds from medical devices) as the most annoying noise sources and the NICU environment regarding noise as "slightly uncomfortable" (41.1%). Since environmental modifications might effectively decrease noise levels [32–34], a study testing the effectiveness of those modifications is proposed in this project, as shown in **Figure 1**.

The data gathered until now is still under analysis, but **Table 5** shows the average noise levels by NICU (the noise levels inside incubators were not considered in this analysis).

NICU	LAeq (dB) Mean (min-max)
A	59.0 (48.3–82.5)
B	52.4 (38.9–71.3)
C	55.8 (42.8–72.8)

Table 5. Average noise levels by NICU.

The noise levels in the three NICUs are higher than that recommended by WHO, which proposes that the average background noise in hospitals should not exceed 35 dB LAeq for areas where patients are treated or observed (**Table 3**). As concentration, precise communication and fast decisions are necessary in the hospital in general, the acoustical environment has to be considered an enormous strain for the staff and a potential risk [35].

5. Conclusion

The main strength of NeoNoise is the combination of strategies to reduce noise levels that are being tested. Furthermore, the different types of studies and approaches, combining questionnaire surveys, educational interventions, and objective measurements provided the collection of a large variety of data, focusing on multiple aspects of staff perception and behavior, as well as factors related to the direct environment of the premature infants. The main concern in NeoNoise was to contribute and to promote healthier environments both for infants and healthcare professionals in hospitals. With that in mind, some particularly important outcomes of this project will be to contribute to educate healthcare staff and to make recommendations to reduce and control noise production in those environments. Health promotion programs should be the mainstream of all interventions and should integrate as much as possible, staff, patients, and visitors. Some limitations of the study are related to the challenge of working in an environment such a NICU, where the tasks and activities performed, are continuously changing due to the evolution of the infants health state.

This work outlines the study design and methods that might be followed by future researchers conducting field studies regarding noise reduction in healthcare facilities. The preliminary findings are relevant to characterize noise exposure of premature infants and staff in NICU. So far, preliminary data analysis revealed that noise levels in the three NICU demonstrated to be higher than recommended. The next step in ongoing analysis is to develop and implement a quiet time protocol, assess its effectiveness and to produce a good practices guide to reduce noise production in a daily basis, improving work conditions as well.

Acknowledgements

The authors would like to thank the assistance given by the Occupational Health Services of the Hospitals involved, as well as to the responsibles of the NICUs under study. The authors also thank to the undergraduate Environmental Health students Mara Nóia, Maria João Coelho, António Azevedo, Raquel Martins, Raquel Lázaro, and Cláudia Nunes (from the School of Allied Health Technologies of Polytechnic Institute of Porto (ESTSP.IPP)) for partial data collection and their support.

Author details

Carlos Carvalhais*, Manuela V. Silva, Ana Xavier and Joana Santos

*Address all correspondence to: caa@estsp.ipp.pt

Environmental Health Department and Research Center on Health and Environment (SOA/CISA), School of Allied Health Technologies of Polytechnic Institute of Porto (ESTSP.IPP), Vila Nova de Gaia, Portugal

References

[1] Berglund B, Thomas L, Dietrich HS. Guidelines for Community Noise [Internet]. Guidelines for Community Noise. Geneva, Switzerland; 1999. Available from: http://www.who.int/docstore/peh/noise/guidelines2.html

[2] Carvalhais C, Santos J, Silva MV, Xavier A. Is it enough training healthcare staff on noise reduction in NIC Units? A pilot study from NeoNoise Project. In: Silva M V., Oliveira R, Rodrigues M, Nunes M, Santos J, Carvalhais C, et al., editors. 3rd International Congress of Environmental Health: Proceedings Book. Porto: Scientific Area of Environmental Health of Allied Health Sciences School of Polytechnic Institute of Porto; Scientific Area of Environmental Health of Allied Health Sciences School of Polytechnic Institute of Porto, Porto, 2014. pp. 220–1.

[3] Santos J, Carvalhais C, Nóia M, Vieira da Silva M, Santos Baptista J. NeoNoise project: Preliminary results of sound pressure levels in a neonatal intensive care unit. In: Arezes P, Baptista JS, Barroso M, Carneiro P, Cordeiro P, Costa N, et al., editors. SHO 2014—International Symposium on Occupational Safety and Hygiene. Guimarães, Portugal: SPOSHO; 2014. pp. 189–93.

[4] Short AE, Short KT, Holdgate A, Ahern N, Morris J. Noise levels in an Australian emergency department. Australas Emerg Nurs J. 2011;14:26–31.

[5] Pugh RJ, Griffiths R. Noise in critical care. Care Crit Ill. 2007;23:105–9.

[6] Carvalhais C, Santos J, Vieira da Silva M, Xavier A. Is there sufficient training of healthcare staff on noise reduction in Neonatal Intensive Care Units? A Pilot Study from NeoNoise Project. J Toxicol Environ Heal Part A. 2015;78:897–903.

[7] Gurses AP, Carayon P. Exploring performance obstacles of intensive care nurses. Appl Ergon. 2009;40:509–18.

[8] Sampaio Neto RA, Mesquita FOS, Paiva Junior MDS, Ramos FF, Andrade FMD, Correia Jr MAV. Noise in the intensive care unit: quantification and perception by healthcare professionals. Rev Bras Ter Intensiva. 2010;22:369–74.

[9] Nunes C, Santos J, da Silva MV, Lourenço I, Carvalhais C. Comparison of different methods for work accidents investigation in hospitals: A Portuguese case study. Work A J Prev Assess Rehabil. 2015;51:601–9.

[10] Mahmood A, Chaudhury H, Valente M. Nurses' perceptions of how physical environment affects medication errors in acute care settings. Appl Nurs Res. 2011;24:229–37.

[11] Ceylan N, Kaba S, Karaman K, Celiker M, Basbugan Y, Demir N. Investigation of the effect of the efficiency of noise at different intensities on the DNA of the newborns. Noise Health. 2016;18:7–9.

[12] Wachman EM, Lahav A. The effects of noise on preterm infants in the NICU. Arch Dis Child Fetal Neonatal Ed. 2011;96:F305–9.

[13] Nicolau AS, Casal D, Lopes PM, Kronenberg P. O The noise in neonatal units of Lisbon and Tejo Valley. Acta Pediatr Port. 2005;36:15–21.

[14] Hammer MS, Swinburn TK, Neitzel RL. Environmental noise pollution in the United States: Developing an effective public health response. Environ Health Perspect. 2014;122:115–9.

[15] Konkani A, Oakley B. Noise in hospital intensive care units-a critical review of a critical topic. J Crit Care. 2012;27:522.e1–522.e9.

[16] Dube J a O, Barth MM, Cmiel C a, Cutshall SM, Olson SM, Sulla SJ, et al. Environmental noise sources and interventions to minimize them: A tale of 2 hospitals. J Nurs Care Qual. 2008;23:216–24.

[17] Connor A, Ortiz E. Staff solutions for noise reduction in the workplace. Perm J. 2009;13:23–7.

[18] Carvalhais C, Santos J, Coelho MJ, Xavier A, Silva M V. Health care staff perception of noise in neonatal intensive care units: A questionnaire survey from NeoNoise Project. In: Arezes P, Baptista JS, Barroso M, Carneiro P, Cordeiro P, Costa N, et al., editors. Occupational Safety and Hygiene IV. London: CRC Press, Taylors & Francis; 2016. pp. 567–71.

[19] Santos J, Miguel AS. Noise levels in hospital environment—The case of intensive care units. In: Arezes P, Baptista JS, Barroso M, Carneiro P, Cordeiro P, Costa N, et al., editors. SHO 2012—International Symposium on Occupational Safety and Hygiene. Guimarães, Portugal: Portuguese Society for Occupational Safety and Hygiene; 2012. pp. 394–6.

[20] Robertson A, Kohn J, Vos P, Cooper-Peel C. Establishing a noise measurement protocol for neonatal intensive care units. J Perinatol Off J Calif Perinat Assoc. 1998;18:126–30.

[21] Wiese CH, Wang LM, Ronsse LM. Comparison of noise levels between four hospital wings with different material treatments. J Acoust Soc Am. 2009;126:2217.

[22] Gray L, Philbin MK. Measuring sound in hospital nurseries. J Perinatol. 2000;20:S100–4.

[23] Kent WT, Tan AW, Clarke MC, Bardell T. Excessive noise levels in the neonatal ICU: potential effects on auditory system development. J Otolaryngol. 2002;31:355–60.

[24] U.S. Environmental Protection Agency. Information on Levels of Environmental Noise Requisite to Protect Public Health and Welfare with an Adequate Margin of Safety. 550/9-74-004. Washington, D.C.: Office of Noise Abatement and Control; 1974. 242 p.

[25] American Academy of Pediatrics: Committee on Environmental Health. Noise: A hazard for the fetus and newborn. Pediatrics. 1997;100:724–7.

[26] Moher D, Shamseer L, Clarke M, Ghersi D, Liberati A, Petticrew M, et al. Preferred reporting items for systematic review and meta-analysis protocols (PRISMA-P) 2015 statement. Syst Rev. 2015;4:1.

[27] Miguel AS, Arezes P, Baptista JS, Melo R, Cordeiro P, Lourenço K, et al. Development and validation of a Guide for the Diagnosis of Occupational Safety and Health Condi-

tions in Local Administration. SPOSHO (Portuguese Society for Occupational Safety and Hygiene); Guimarães, 2010. 91 p.

[28] Christensen M. What knowledge do ICU nurses have with regard to the effects of noise exposure in the Intensive Care Unit? Intensive Crit Care Nurs. 2005;21:199–207.

[29] Chen HL, Chen CH, Wu CC, Huang HJ, Wang TM, Hsu CC. The influence of neonatal intensive care unit design on sound level. Pediatr Neonatol [Internet]. 2009;50:270–4.

[30] Machado MC, Alves MI, Couceiro ML. Child and Youth Health in Portugal: National Health Plan indicators. Acta Pediátrica Port. 2011;42:195–204.

[31] Bremmer P, Byers JF, Kiehl E. Noise and the premature infant: physiological effects and practice implications. J Obstet Gynecol Neonatal Nurs. 2003;32:447–54.

[32] Philbin MK, Gray L. Changing levels of quiet in an intensive care nursery. J Perinatol. 2002;22:455–60.

[33] Philbin MK, Klaas P. Evaluating studies of the behavioral effects of sound on newborns. J Perinatol. 2000;20:S61–7.

[34] Ramesh A, Rao PNS, Sandeep G, Nagapoornima M, Srilakshmi V, Dominic M, et al. Efficacy of a low cost protocol in reducing noise levels in the neonatal intensive care unit. Indian J Pediatr. 2009;76:475–8.

[35] Siegmann S, Notbohm G. Noise in hospitals as a strain for the medical staff. Proceedings of Meetings on Acoustics Meeting Acoustics. Acoustical Society of America: Montreal, 2013. pp. 040092–040092.